BASIC fracture mechanics

including an introduction to fatigue

R N L Smith, BSc, MSc, PhD, FIMA

Senior Lecturer
Applied and Computational Mathematics Group
RMCS (Cranfield)
Shrivenham
Swindon
Wilts

To
Barbara and Les

Butterworth-Heinemann Ltd
Halley Court, Jordan Hill, Oxford OX2 8EJ

PART OF REED INTERNATIONAL BOOKS

OXFORD LONDON BOSTON MUNICH
NEW DELHI SINGAPORE SYDNEY
TOKYO TORONTO WELLINGTON

First published 1991

© Butterworth-Heinemann Ltd 1991

All rights reserved. No part of this publication may be reproduced in
any material form (including photocopying or storing in any medium
by electronic means and whether or not transiently or incidentally to
some other use of this publication) without the written permission of
the copyright holder except in accordance with the provisions of the
Copyright, Designs and Patents Act 1988 or under the terms of a
licence issued by the Copyright Licensing Agency Ltd, 33–34 Alfred
Place, London WC1E 7DP, England. Applications for the copyright
holder's written permission to reproduce any part of this publication
should be addressed to the publishers

British Library Cataloguing in Publication Data
Smith, R.N.L.
 BASIC fracture mechanics.
 I. Title
 620.100285

ISBN 0 7506 1489 7

Library of Congress Cataloging in Publication Data
Smith, R. N. L.
 Basic fracture mechanics : including an introduction to fatigue /
R.N.L. Smith.
  p. cm.
 Includes index.
 ISBN 0-7506-1489-7 : $9.95
 1. Fracture mechanics—Data processing. 2. Materials—Fatigue–
 –Data processing. I. Title.
 TA409.S63 1991
 620.1'126—dc20   91-7417
                                       CIP

Printed and bound in Great Britain by
Biddles Ltd, Guildford and King's Lynn

Contents

Preface

1 Introduction to BASIC — 1

1.1 The BASIC approach — 1
1.2 The elements of BASIC — 1
1.3 Checking Programs — 9
1.4 Different computers and variants of BASIC — 10
1.5 Summary of BASIC statements — 10
1.6 Bibliography — 11

2 Elastic fracture — 12

2.1 Introduction — 12
2.2 Stress and strain — 13
2.3 Stress concentrations — 18
2.4 Energy and crack growth — 19
2.5 Crack stresses and displacements — 24
2.6 References — 31
PROBLEMS — 31

PROGRAMS
2.4 CRIT: Critical crack size or stress — 22
2.5 CRSS: Crack displacements and stresses — 28

3 Stress intensity factors — 33

3.1 Introduction — 33

3.2 Using tables and graphs — 33
3.3 Superposition — 38
3.4 Numerical methods — 43
3.5 Using stresses and displacements — 50
3.6 Energy relationships — 55
3.7 Green's functions — 58
3.8 Weight functions — 61
3.9 References — 66
PROBLEMS — 66

PROGRAMS
3.2 LAG: Lagrange interpolation — 35
3.3 MS: K-values from mean/max/tip stress — 41
3.5 KDISP: K_I, K_{II} from nodal displacements — 51
3.7 WFI: K by integrating a Green's function — 60
3.8 WFEDG: K by integrating a weight function — 64

4 Crack tip plasticity — 69

4.1 Introduction — 69
4.2 Plastic zone size — 70
4.3 The shape of the plastic zone — 74
4.4 Elastic-plastic failure — 78
4.5 Plane strain fracture toughness — 80
4.6 Failure at higher levels of plasticity — 88
4.7 References — 91
PROBLEMS — 91

PROGRAMS
4.2 DUGIN: Dugdale/Irwin plastic zone size — 73
4.3 ZONE: shape of plastic zone — 76
4.5 CTS: Compact tension specimen for fracture toughness — 83
4.5.1 DCTS: Design of the compact tension specimen — 86

5 Crack growth — 93

5.1 Introduction — 93
5.2 Fatigue crack growth — 94
5.3 Second order terms — 101

5.4 Predicting the direction of crack growth	105
5.5 Fast fracture under mixed-mode loading	113
5.6 Fatigue under mixed-mode loading	116
5.7 References	118
PROBLEMS	118

PROGRAMS

5.2 PCG: Crack growth using the Paris law	96
5.3 CRSSECO: Second order crack stress and displacement	103
5.4 SED: Strain energy density and maximum tensile stress	107

6 Some applications in fracture mechanics — 120

6.1 Introduction	120
6.2 Design strategies	121
6.3 The critical crack length	123
6.4 Residual stresses	126
6.5 Notches and fatigue crack growth	129
6.6 Experimental data	133
6.7 Failure assessment diagrams	138
6.8 Three-dimensional cracks	141
6.9 References	146
PROBLEMS	147

PROGRAMS

6.3 ACRIT: Critical crack length	125
6.5 NOTCH: Minimum stress for notch fatigue	132
6.6 DAN:Gradient of a, N curve	135
6.6.1 LSQ: Least square straight line fit	137
6.8 ELL: K-values for a semi-elliptic surface crack	145

Preface

Fracture mechanics is becoming increasingly important to engineering design as designers search for ways of producing ever lighter structures. Lightweight designs are clearly crucial in the aerospace industry but are becoming of much wider significance in the pursuit of greater efficiency in almost all branches of engineering. Unfortunately the high strength materials required are prone to develop cracks which may grow slowly under quite modest stresses until there is a sudden catastrophic failure.

The aim of this book is to introduce the reader to fracture and fatigue and to provide a series of programs which implement those methods which are both useful and simple. Many of the methods of fracture mechanics which are, in principle, quite simple, may in fact be obscured by the often tedious or complex hand calculations required. The topic is well suited to a computer based approach since repetitive calculations can be done very quickly and simply with just a few lines of BASIC. The reader has at the simplest level a series of 'black boxes' which allows the study of say the effects of crack length on failure stress, without an intimate understanding of the manipulation of algebraic formulae. Of course, all the formulae used in the programs are explained or derived in the text for more thorough readers.

This book uses programs written in BASIC because it is probably the most widely available and easily understood computer language. Almost any computer can be used ranging from large 'Mainframe' machines to the personal computer and even some 'pocket calculators'. BASIC is probably the most straight-forward computer language to learn and students can be operating 'hands on' writing very simple programs in their first study period. No special compiling, linking or editing commands need to be learnt before starting on the language itself, although compiled versions may be available on some machines.

One of the drawbacks of BASIC lies in its poor subroutine facilities in standard form (some improved versions are available) but the programs given here would be virtually of the same form in many languages. This is because the programs are written not to demonstrate programming skill or for computational efficiency, but in order to calculate in a way that relates simply to the text. An experienced programmer could translate the programs into say, FORTRAN while reading the BASIC, and may well add more comments and spacing since these are minimised here.

Chapter 1 is, as is usual with this series, a brief review of the BASIC language, while chapter 2 revises some useful formulae in stress analysis and continues to discuss stress concentrations. This leads to the idea of infinite stress at the crack tip, the stress intensity factor and the energy required for crack growth. Chapter 3 reviews some methods of obtaining stress intensity factors including the use of results obtained using finite or boundary elements although the actual FEM and BEM methods are not given. Having thus far assumed purely linear elasticity, in chapter 4 we outline some models of plastic behaviour in the highly-stressed crack tip region which leads naturally to the discussion of fatigue growth in chapter 5. Also in chapter 5 we introduce the idea of second-order fracture parameters and consider their effects on the initiation and direction of crack growth. Chapter 6 introduces the ideas of fracture-conscious design and inspection which aims to avoid cracks growing to near-failure lengths. We discuss the problems of residual stresses, crack growth from notches and experimentally derived fatigue results. Failure assessment diagrams which aim to incorporate fracture mechanics and plastic collapse are briefly illustrated. Finally, we discuss the use of numerical results for three-dimensional fracture problems in the context of the elliptical crack.

I gratefully acknowledgement the help of a number of people who have read and criticised part or all of the manuscript. These include Mike Iremonger and Peter Smith, the series editors, Mike Edwards of Materials group RMCS and David Rooke of the Royal Aerospace Establishment at Farnborough. I must also thank my wife Lynne and son Ben for their forbearance !

R.N.L.S.

Chapter 1

Introduction to BASIC

1.1 The BASIC approach

The programs in this book are written in the BASIC programming language. BASIC, an acronym for Beginner's All-purpose Symbolic Instruction Code, was developed at Dartmouth College, USA as an easy-to-learn, general-purpose language. Originally intended for use on timesharing computer systems, it has gained widespread popularity as the main language associated with microcomputers. Not only is the language easy to learn but it is also particularly easy to use. Without complication a program can be written, typed in at the computer, run and corrected and run again if any errors are present. The main disadvantages of simple BASIC relate to its lack of structure (see Section 1.4) but this is not an important consideration for short programs such as those in the following chapters.

This book aims to help in the learning of BASIC by applying it to a relevant engineering subject. This aim can be met by the reader studying the examples, possibly copying them and then trying some of the problems. Although this book does not specifically teach the grammar of BASIC, a short description of the simple BASIC used is given in the next section.

1.2 The elements of BASIC

1.2.1 Mathematical expressions

One of the main objects of the example programs in this book is to evaluate the equations that arise in fracture mechanics. These equations contain numerical constants, variables (e.g. x) and functions (e.g. sine(x)). All numbers are treated identically whether they are integer (e.g. 36) or real (e.g. 36.1). An exponential form is used to represent large or small numbers (3.61×10^6 is written $3.61E6$). Numeric variables are represented by a letter or a letter followed by a digit (e.g. E or E1). On many computers π is directly available to the user either as PI or as a π key. For generality π is always computed within the program as $4tan^{-1}(1.0)$ in this book. An operation, such as square root, can be done using an in-built function (SQR(X)). The argument in brackets (X) can be a number, a variable or a mathematical expression. For trigonometric functions (SIN(X), COS (X), etc.) the argument is interpreted as being measured in radians. Other functions include a natural logarithm and its exponential (LOG and EXP respectively), ABS which selects the absolute value of an argument and INT which selects the integer part of an argument.

Mathematical equations also contain operators such as plus and minus, etc. These operations have a hierarchy in that some are performed by the computer before others. In descending order of hierarchy the operators are

```
to the power of (^)
multiply (*) and divide (/)
add (+) and subtract (-)
```

Thus, for example, multiplication is done before addition. The computer works from left to right if the operators have the same hierarchy. Brackets can be used to override any of these operations. Hence (a+b)/3c becomes (A + B)/(3*C) or (A + B)/3/C.

1.2.2 Program structure and assignment

A BASIC program is a sequence of statements which define a procedure for the computer to follow. As it follows this procedure the computer allocates values to each of the variables. The values of some of these variables may be specified by data that is input to the program. Others are generated in the program using, for instance, the assignment statement. This has the form

line number [LET] variable = mathematical expression

where the word LET is usually optional and therefore omitted. For example one of the roots of a quadratic equation.

$$x1 = \frac{-b + \sqrt{b^2 - 4ac}}{2a}$$

may be obtained from a statement such as

    100 X1 = (-B + SQR(B^2-4*A*C))/(2*A)

It is important to realise that an assignment statement is not itself an equation. It is an instruction to give the variable on the left-hand side the numeric value of the expression on the right-hand side. Thus it is possible to have a statement

    50 X = X+1

which increases the value of X by 1.

Each variable can have only one value at any time unless it is subscripted (see section 1.2.7).

Note that all BASIC statements (i.e. all the program lines) are numbered. This defines the order in which they are executed.

1.2.3 Input

For 'interactive conversational' programs the user specifies variables by 'run-time' input of their values. The statement has the form

    line number INPUT variable [,variable 2,...]

e.g.

    20 INPUT A, B, C

When the program is run the computer prints ? as it reaches this statement and waits for the user to type values for the variables, e.g.

    ? 5, 10, 15

which makes A = 5, B = 10 and C = 15 in the above example.

Most computers will also allow a string in quotes as part of the INPUT statement. This string is printed at run-time and can thus be used to clarify which parameter value is required. The statement

    20 INPUT "WHAT ARE A,B,C";A, B, C

prints the query

    WHAT ARE A,B,C ?

when the program runs and the values are then typed in as above.

There is an alternative form of data input which is useful if there are a lot of data or if the data are not to be changed by the user. For this type of data specification there is a statement of the form

   line number READ variable 1 [,variable 2,...]

e.g.

      20 READ A,B,C

with an associated statement (or number of statements) of the form

   line number DATA number 1 [,number 2.,,,]

e.g.

      1 DATA 5,10,15.6

or

      1 DATA 5
      2 DATA 10
      3 DATA 15.6

DATA statements can be placed anywhere in a program - it is often convenient to place them at the beginning or the end so they can be easily changed.

When using built-in data it is sometimes necessary to read the data from their start more than once during a single program run. This is done using the statement

   line number RESTORE

1.2.4 Output

Output of data and the results of calculations, etc. is done using a statement of the form

   line number PRINT list

This list may contain variables or expressions, e.g.

      200 PRINT A,B,C,A*B/C

text enclosed in quotes, e.g.

```
10 PRINT "INPUT A,B,C IN MM";
```

or mixed text and variables, e.g.

```
300 PRINT "STRESS IS";S;"N/MM^2"
```

The items in the list are separated by commas or semi-colons. Commas give tabulation in columns, each about 15 spaces wide. A semi-colon suppresses this spacing and if it is placed at the end of a list it suppresses the line feed. If the list is left unfilled a blank line is printed.

PRINT statements may also be used in association with both 'runtime' input (to indicate what input is required) and READ/DATA statements (because otherwise the program user has no record of the data).

1.2.5 Conditional statements

It is often necessary to enable a program to take some action, if, and only if, some condition is fulfilled. This is done with a statement of the form

line number IF expression 1 (operator) expression 2 THEN line number

where the possible conditional operators are

```
 =  equals
 <> not equal to
 <  less than
 <= less than or equal to
 >  greater than
 >= greater or equal to
```

For example a program could contain the following statements if it is to stop when a zero value of A is input, i.e.

```
20 INPUT A
30 IF A <> 0 THEN 50
40 STOP
50 ...
```

Note the statement

  line number STOP

which stops the run of a program.

1.2.6 Loops

There are several means by which a program can repeat some of its procedure; the self-repeating sequence of program statement is called a loop. The simplest such statement is

  line number GO TO line number

This can be used, for instance, with the above conditional statement example so that the program continues to request values of A until the user inputs zero.

The most common means of performing loops is with a starting statement of the form

  line number FOR variable = expression 1 TO expression 2 [STEP expression 3]

where the STEP is assumed to be unity if omitted. The finish of the loop is signified by a statement

  line number NEXT variable

where the same variable is used in both FOR and NEXT statements. Its value should not be changed in the intervening lines.

A loop is used if, for example, N sets of data have to be READ and their reciprocals printed, e.g.

```
10 READ N
20 PRINT "NUMBER","RECIPROCAL"
30 FOR I=1 TO N
40 READ A
50 PRINT A,1/A
60 NEXT I
```

Loops can also be used to generate data. Consider the simple temperature conversion program

```
10 PRINT "CENTIGRADE","FAHRENHEIT"
20 FOR C = 0 TO 100 STEP 5
30 PRINT C,9*C/5+32
40 NEXT C
```

1.2.7 Subscripted variables

It is sometimes very beneficial to allow a single variable to have a number of different values during a single program run. For instance, if a program contains data for several materials it is convenient for their densities to be called R(1), R(2), R(3), etc. instead of R1, R2, R3, etc. It is then possible for a single statement to perform calculations for all the materials, e.g.

```
50 FOR I = 1 TO N
60 M(I) = V*R(I)
70 NEXT I
```

which determines the mass M(I) for each of the N materials from the volume (V) of the body.

A non-subscripted variable has a single value associated with it and if a subscript variable is used it is necessary to provide space for all the values. This is done with a dimensioning statement of the form.

line number DIM variable 1 (integer 1) [,variable 2 (integer 2), ...]

e.g.

```
20 DIM R(50),M(50)
```

which allows up to 50 values of R and M. The DIM statement must occur before the subscripted variables are first used.

On some computers it is possible to use a dimension statement of a different form, e.g.

```
20 DIM R(N),M(N)
```

where the value of N has been previously defined. This form, when available, has the advantage of not wasting space.

1.2.8 Subroutines

Sometimes a sequence of statements needs to be accessed more than once in the same program. Instead of merely repeating these statements it is better to put them in a subroutine. The program then contains statements of the form

> line number GOSUB line number

When the program reaches this statement it branches (i.e. transfers control) to the second line number. The sequence of statements starting with this second line number ends with a statement

> line number RETURN

and the program returns control to the statement immediately after the GOSUB call.

Subroutines can be placed anywhere in the program but it is usually convenient to position them at the end, separate from the main program statements.

Another reason for using a subroutine occurs when a procedure is written which is required in more than one program. It is often desirable to use less common variable names (e.g. X9 instead of X) in such subroutines. This minimises the possibility of the same variable name being used with a different meaning in separate parts of a program.

1.2.9 Other statements

(1) Explanatory remarks or headings which are not to be output can be inserted into a program using

> line number REM comment

Any statement beginning with the word REM is ignored by the computer. On some computers it is possible to include remarks on the same line as other statements.

(2) Non-numeric data (e.g. words) can be handled by string variables. A string is a series of characters within quotes, e.g. 'STRESS' and a string variable is a letter followed by a $ sign, e.g. S$. They are particularly valuable when printed headings need to be varied.

(3) Multiple branching can be done with statements of the form

line number ON expression THEN line number 1 [,line number 2,...]

and

line number ON expression GOSUB line number 1 [, line number 2,...]

When a program reaches one of these statements it branches to line number 1 if the integer value of the expression is 1, to line number 2 if the expression is 2, and so on. An error message is printed if the expression gives a value less than 1 or greater than the number of referenced line numbers.

(4) Functions (other than those built into the language such as SIN(X)) can be created as defined functions using a DEF statement. For example

```
10 DEF FNA(X) = X^3 + X^2 + X + 1
```

defines a cubic function which can be recalled later in the program as FNA (variable) where the value of this variable is substituted for X. A defined function is useful if an algebraic expression is to be evaluated several times in a program.

1.3 Checking Programs

Most computers give a clear indication if there are grammatical (syntax) errors in a BASIC program. Program statements can be modified by retyping them correctly or by using special editing procedures. The majority of syntax errors are easy to locate but if a variable has been used with two (or more) different meanings in separate parts of the program some mystifying errors can result.

It is not sufficient for the program to be just grammatically correct. It must give the correct answers. A program should therefore be checked either by using data which give a known solution or by hand calculation. If the program is to be used with a wide range of data or by users other than the program writer, it is necessary to check that all parts of it function. It is also important to ensure that the program does not give incorrect but plausible answers when 'nonsense' data are input. It is quite difficult to make programs completely 'userproof' and they become lengthy in so doing. The programs in this been kept as short as possible for the purpose of clarity and may not therefore be fully 'userproof'.

1.4 Different computers and variants of BASIC

The examples in this book use a simple version of BASIC that should work on most computers, even those with a small storage capacity. Only single-line statements have been used though many computers allow a number of statements on each line with a separator such as \. Multiple assignments may also be possible so that, for instance, five identical assignments for different variables can be replaced with a single line such as

$$1040 \; S1 = S2 = S3 = S4 = S5 = 0$$

There is one important feature which distinguishes computers, particularly microcomputers, with a visual display unit (VDU). This concerns the number of available columns across each line and the number of rows that are visible on the screen. Simple modifications of some of the programs may be necessary to fit the output to a particular microcomputer. TAB printing is a useful facility for this purpose.

Various enhancements to BASIC have been made since its inception and these are implemented on a number of computer systems. The programs in this book could be re-written to take account of some of these 'advanced' features. For example, the ability to use long variable names (e.g. STRESS instead of, say, S or S1) makes it easier to write unambiguous programs. Other advanced facilities include more powerful looping and conditional statements and independent subroutines which make the writing of structured programs easier. In simplistic terms, structured programming involves the compartmentalisation of programs and minimises branching due to statements containing 'GO TO line number' and 'THEN line number'. Good program structure is advantageous for long programs.

1.5 Summary of BASIC statements

```
Assignment
    LET            Computes and assigns value
    DIM            Allocates space for subscripted variables
```

Summary of BASIC statements

Input
- INPUT — 'Run-time' input of data
- READ — Reads data from DATA statements
- DATA — Storage area for data
- RESTORE — Restores DATA to its start

Output
- PRINT — Prints output

Program control
- STOP — Stops programs run
- GO TO — Unconditional branching
- IF ... THEN — Conditional branching
- FOR ..TO..STEP — Opens loop
- NEXT — Closes loop
- GOSUB — Transfers control to subroutine
- RETURN — Returns control from subroutine
- ON...THEN — Multiple branching
- ON-GOSUB — Multiple subroutine transfer

Comment
- REM — Comment in program

Functions
- SQR — Square root
- SIN — Sine (angle in radians)
- COS — Cosine (angle in radians)
- ATN — Arctangent (gives angle in radians)
- LOG — Natural logarithm (base e)
- EXP — Exponential
- ABS — Absolute value
- INT — Integral value
- DEF FN — Define function

1.6 Bibliography

Alcock, D., *Illustrating BASIC*, Cambridge University Press, (1977).
Kemeny, J.G. and Kurtz, T.E., *BASIC Programming*, Wiley, (1968).
Monro, D.M., *Interactive Computing with BASIC*, Edward Arnold, (1974).

Chapter 2

Elastic Fracture

2.1 Introduction

The in-service failure of a component may have catastrophic consequences. Breakages occur because most materials contain large numbers of small, perhaps undetectable, cracks any one of which may grow under the stresses of use until it spreads rapidly and disastrously across the whole component.

Since different materials behave in very different ways when a load is applied it is useful to consider possible extremes of behaviour. When a piece of plasticine is stretched and then released, it does not return to its original shape but is permanently deformed. If the plasticine is stretched further, it continues to deform, becoming thinner and thinner, until it finally breaks near the thinnest point. This process is known as ductile failure and is always associated with permanent deformation. In contrast, a sheet of glass breaks due only to crack growth and the pieces may be reassembled into a flat sheet showing that there is no significant permanent deformation - a brittle failure. Common engineering materials lie somewhere between these two extremes and their nature may change with the conditions of use. For example, mild steel may behave in a brittle fashion in arctic temperatures and as a ductile material in more temperate regions. In this book we are concerned with predominantly brittle failure which is the domain of the science of Fracture Mechanics.

Lightweight designs are required for a broad range of engineering applications and by their very nature imply the use of high strength materials. Aircraft structures are one example, typically utilising high strength alloys of steel, aluminium or titanium which usually fail in a predominantly brittle fashion due to crack growth. The aim of the designer is to ensure that failure does not occur in service. He should be able to estimate the size of the largest crack which precludes failure and determine inspection intervals so that a specified smaller crack cannot grow to critical size between inspections. The task is complicated by the theoretical nature of crack problems which imposes an additional level of difficulty in the mathematics of stress analysis. This additional complexity means that few analytical solutions are available and the fracture-conscious designer must usually resort to numerical methods or other approximations to produce safe structures.

To analyse a realistic model of a real material is extremely difficult and in this chapter we discuss the simplifications necessary to produce a useful theoretical model. After a brief review of some useful elasticity formulae, we then examine the properties of a single isolated crack tip.

2.2 Stress and Strain

This section is intended as a brief review of some important and useful formulae - not as a comprehensive introduction to elasticity! For a simple introduction with some useful programs see Iremonger (1982).

Direct Stress

If a linearly elastic bar length L, cross-sectional area A, is stretched by a force P and extends by an amount δ (figure 2.1), then:

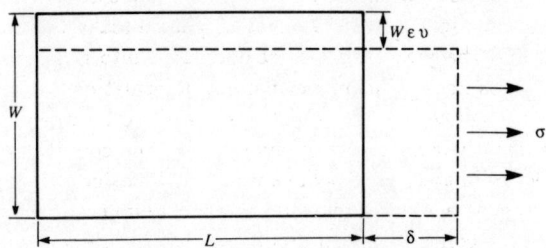

Figure 2.1 Direct stress and strain

(a) *Stress* $\sigma = P/A$
(b) *Strain* $\epsilon = \delta/L$
(c) *Hookes Law* $\sigma = E\epsilon$ where E is *Youngs Modulus* .
(d) *Axial strain* $= -\nu\epsilon$ where ν is *Poisson's ratio*.

Shear Stress

If a force Q is applied to the sides of area A of the square cross-section shown in figure 2.2 then:

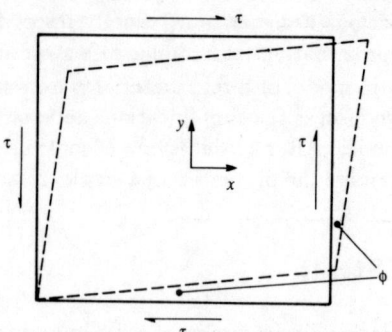

Figure 2.2 Shear stress

(a) *Shear stress* $\tau_{xy} = Q/A$
(b) *Shear strain* $\epsilon_{xy} = \frac{1}{2}\tan\phi$ ('engineering' shear strain $\gamma = 2\epsilon_{xy}$)
(c) *Hookes Law* $\tau_{xy} = 2\mu\epsilon_{xy}$ where μ is the *Shear Modulus* or modulus of rigidity.
(d) *Shear Modulus* $\mu = E/[2(1+\nu)]$

Three-dimensional stress and strain

If the elastic material is not of uniform cross-section, then the above definitions of stress and strain must be replaced by the limiting values (over a vanishingly small area) of the appropriate differentials. In three-dimensions, there are three components of direct strain ϵ_x, ϵ_y, ϵ_z with corresponding shear strains. Thus, if u is the displacement in the x direction and v is the displacement in the y direction then

$$\epsilon_x = \partial u/\partial x, \quad \epsilon_y = \partial v/\partial y, \quad \epsilon_{xy} = \frac{1}{2}(\partial v/\partial x + \partial u/\partial y) \qquad (2.1)$$

The nine components of stress are illustrated in figure 2.3. Strictly speaking it is correct to write all stresses and strains with two subscripts since they are tensor quantities. However, we use the common short forms $\epsilon_x \equiv \epsilon_{xx}$, $\sigma_x \equiv \tau_{xx}$ etc. Note that in order to preserve

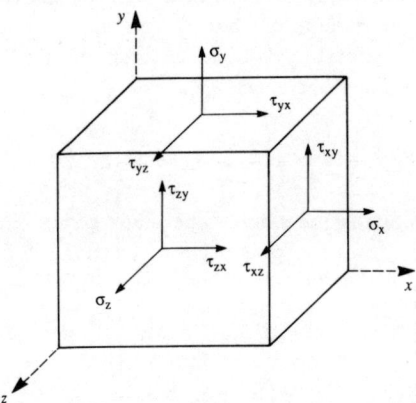

Figure 2.3 Three-dimensional stresses

equilibrium we require:
$$\tau_{zx} = \tau_{xz}, \quad \tau_{zy} = \tau_{yz}, \quad \tau_{yx} = \tau_{xy} \qquad (2.2)$$

The full three-dimensional version of Hooke's law is:
$$\epsilon_x = \frac{1}{E}[\sigma_x - \nu(\sigma_y + \sigma_z)], \qquad \epsilon_{yz} = \frac{1+\nu}{E}\tau_{yz}$$
$$\epsilon_y = \frac{1}{E}[\sigma_y - \nu(\sigma_x + \sigma_z)], \qquad \epsilon_{xz} = \frac{1+\nu}{E}\tau_{xz}$$
$$\epsilon_z = \frac{1}{E}[\sigma_z - \nu(\sigma_x + \sigma_y)], \qquad \epsilon_{xy} = \frac{1+\nu}{E}\tau_{xy} \qquad (2.3)$$

However, a fully three-dimensional analysis of a problem is usually rather difficult, and therefore a two-dimensional approximation is often used.

Plane stress assumes that the stress through the thickness is negligible compared to other stresses, ie that $\sigma_z = \tau_{zx} = \tau_{zy} = 0$. This is approximately true in a thin sheet of material. The simplified Hooke's law becomes:
$$\epsilon_x = \frac{1}{E}(\sigma_x - \nu\sigma_y), \quad \epsilon_y = \frac{1}{E}(\sigma_y - \nu\sigma_x), \qquad (2.4)$$

with ϵ_{xy} remaining unchanged. If the strain in the z-direction is required it is calculated from:
$$\epsilon_z = -\frac{\nu}{E}(\sigma_x + \sigma_y) \qquad (2.5)$$

Plane strain assumes that the strain through the thickness is negligible compared to other strains or that $\epsilon_z = \epsilon_{zx} = \epsilon_{zy} = 0$. This condition is approximately true in a thick sheet of material. The simplified Hooke's law becomes:

$$\epsilon_x = \frac{1-\nu^2}{E}\left(\sigma_x - \frac{\nu}{1-\nu}\sigma_y\right), \quad \epsilon_y = \frac{1-\nu^2}{E}\left(\sigma_y - \frac{\nu}{1-\nu}\sigma_x\right) \quad (2.6)$$

and again ϵ_{xy} remains the same. The plane strain equations may be obtained directly from the plane stress equations if E is replaced by $E/(1-\nu^2)$ and ν by $\nu/(1-\nu)$. The stresses in plane strain are:

$$\tau_{xz} = \tau_{yz} = 0$$

$$\sigma_z = \nu(\sigma_x + \sigma_y) \quad (2.7)$$

Stress transformation

It is often useful to resolve stresses in a particular direction, to take advantage of symmetry for example. If we require the stresses on a

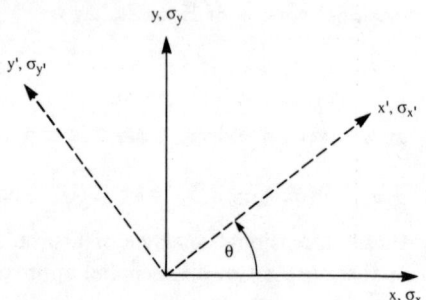

Figure 2.4 Stress transformation

new set of axes at an angle θ to the old ones as in figure 2.4 then:

$$\sigma_{x^1} = \frac{\sigma_x + \sigma_y}{2} - \left(\frac{\sigma_y - \sigma_x}{2}\right)\cos 2\theta + \tau_{xy}\sin 2\theta$$

$$\sigma_{y^1} = \frac{\sigma_x + \sigma_y}{2} + \left(\frac{\sigma_y - \sigma_x}{2}\right)\cos 2\theta - \tau_{xy}\sin 2\theta$$

$$\tau_{x^1y^1} = \tau_{xy}\cos 2\theta + \left(\frac{\sigma_y - \sigma_x}{2}\right)\sin 2\theta \quad (2.8)$$

Stress and Strain

If the shear stress in the new axes is chosen to be zero, then a solution to equations 2.8 exists for two values of theta at right angles to each other - the *principal directions*. The direct stresses in these directions are the *principal stresses* which are given by:

$$\sigma_1 = \frac{\sigma_x + \sigma_y}{2} + \left[\left(\frac{\sigma_y - \sigma_x}{2}\right)^2 + \tau_{xy}^2\right]^{1/2}$$

$$\sigma_2 = \frac{\sigma_x + \sigma_y}{2} - \left[\left(\frac{\sigma_y - \sigma_x}{2}\right)^2 + \tau_{xy}^2\right]^{1/2} \quad (2.9)$$

For a three-dimensional problem there will be three principal stresses and it is usual to arrange these in descending order so that $\sigma_1 \geq \sigma_2 \geq \sigma_3$. For plane strain problems it is important to include the through-thickness stresses in the principal stresses.

Strain energy

The load-extension graph for a linearly elastic material without a crack is shown in figure 2.5. The *strain energy* U of this system is the work

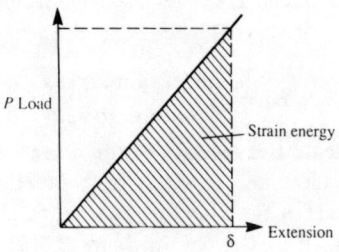

Figure 2.5 load versus extension for an uncracked region

done in producing the extension δ and is the area under the load-extension graph. The area may be obtained directly or by integration giving:

$$U = \int_0^\delta F dx = \int_0^\delta \frac{P}{\delta} x \, dx = \frac{1}{2} P \delta \quad (2.10)$$

Similarly, the strain energy per unit volume U_v for an area A length L is

$$U_v = \frac{1}{2} \frac{P}{A} \frac{\delta}{L} = \frac{1}{2} \sigma \epsilon = \frac{\sigma^2}{2E} \quad (2.11)$$

2.3 Stress concentrations

Analytical solutions are available for some rather special cases in stress analysis. In order to obtain such solutions it is necessary to make some simplifying assumptions about the properties of materials. Materials are usually assumed to be homogeneous and isotropic, ie of uniform composition and physical properties, without grains or grain boundaries and without surface hardening. For the present, materials are also assumed to be perfectly elastic so that they return to their original shape when loads are removed. Very thick or very thin plates may be approximated as two-dimensional regions assuming conditions of plane strain or plane stress respectively. This reduction to two-dimensions greatly simplifies mathematical analysis since complex variable methods may be used. We do not consider the analysis here for simplicity and results are therefore quoted without proof (for a simple introduction see Parker 1981).

The elliptical hole in an infinite elastic sheet under uniform tension (figure 2.6(a)) is one example where stresses can be predicted using mathematical analysis. The effect of the hole is to magnify the applied (or nominal) stress producing a stress concentration which varies around its circumference. The stress concentration factor is then defined as

$$K_t = \frac{maximum\ stress}{nominal\ stress} \quad (2.12)$$

and is used to indicate the severity of the defect. In one special case, when the ellipse is circular, the stress concentration factor is a maximum of 3 at the ends of the diameter normal to the applied stress. A shallower ellipse, such as the one in figure 2.6(a), will have a higher stress concentration at the same points. A complete analysis of this problem reveals that for unit applied stress, the stresses on the end of the ellipse take the form:

$$\sigma = 1 + 2\sqrt{a}/\sqrt{r} \quad (2.13)$$

where r is the radius of the tip of the ellipse and a is the half-length of the ellipse. As r tends to zero, the ellipse degenerates into a perfectly flat crack and the stress is then theoretically infinite at the crack tip. Of course, in real materials this limit is never reached since there will be some deformation at the crack tip. Nevertheless, theoretical results are invaluable in predicting the behaviour of real materials.

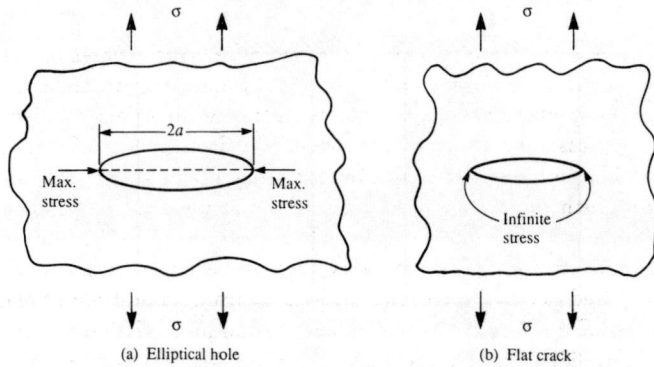

Figure 2.6 Stress concentration in an infinite plate

The flat crack is often drawn as in figure 2.6(b), to make it clear that the crack has two surfaces. However, it is important to realise that theoretically the crack consists of a line discontinuity, like a cut made by a razor blade in a sheet of paper, unless crack opening forces are applied.

Stress concentrations are often responsible for rapid crack growth and knowledge of their effects is vital to fracture prediction. One useful compendium of data is given by Peterson (1974).

2.4 Energy and crack growth

In the previous section we saw that crack tip stresses are theoretically infinite. This might seem to imply that under any opening load a cracked body will break immediately. To understand why this does not happen, we must look at the energy balance between the applied load and the creation of new crack surfaces.

Consider a rectangular plate which is stretched by a line of equal and opposite forces at each end and then clamped in position - figure 2.7(a). If a crack is now formed in the centre of the plate we have a system like that shown in figure 2.7(b) where the opening of the crack has relaxed the system. The strain energy U of the system has now been reduced and for one crack tip, Griffith showed that the amount of energy released is:

$$U = \frac{\sigma^2}{2E}\pi a^2 \qquad (2.14)$$

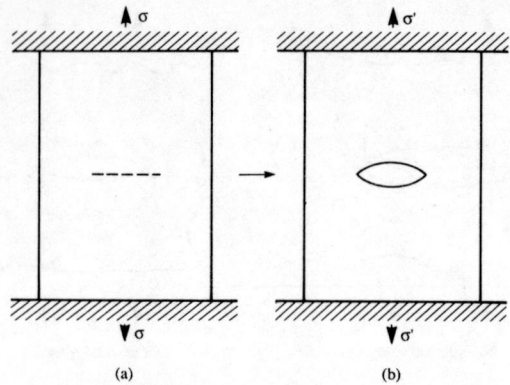

Figure 2.7 Relaxation of a clamped plate

assuming plane stress conditions. For plane strain E is replaced by $E/(1-\nu^2)$ and hence:

$$U = \frac{\sigma^2}{2E}(1-\nu^2)\pi a^2 \tag{2.15}$$

To create new crack surfaces requires energy however, and the surface energy W required is given by

$$W = 2a\gamma \tag{2.16}$$

where γ is the surface energy density or surface tension. A crack will only grow if, by doing so, it reduces the total energy ($= U + W$) of the system. The way in which the energies concerned vary with crack length is shown in figure 2.8. Note that U is negative since it represents a release of energy. For a small crack ($a < a_c$), an increase in length means the system has a higher total energy and this can only be achieved if additional energy is supplied from outside the system. If however, the crack length is greater than the critical value a_c, an increase in length reduces the total energy and the crack will grow unstably. The point a_c is a maximum point and hence the gradient of the total energy is zero. Mathematically:

$$\frac{\partial}{\partial a}(W + U) = \frac{\partial W}{\partial a} + \frac{\partial U}{\partial a} = 0 \tag{2.17}$$

Energy and crack growth

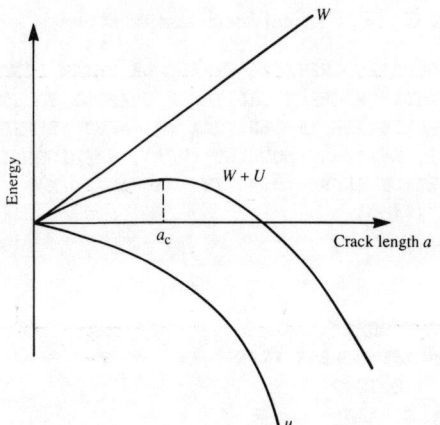

Figure 2.8 Variation of energy with crack length

If we define the *energy release rate G* as

$$G = -\frac{\partial U}{\partial a} = \frac{\pi}{E}\sigma^2 a \qquad (2.18)$$

and the *crack resistance force R* as

$$R = \frac{\partial W}{\partial a} = 2\gamma \qquad (2.19)$$

then equation 2.17 is equivalent to

$$R - G = 0 \qquad (2.20)$$

and we have unstable crack growth if $G > R$ or, using equations 2.18 and 2.19, if $\sigma > \sigma_c$ where

$$\sigma_c = \sqrt{\frac{2E\gamma}{\pi a}} \qquad (2.21)$$

where E becomes $E/(1-\nu^2)$ for plane strain.

Unstable crack growth will normally lead to instant component failure so that cracks which approach the critical condition cannot be tolerated. There are, however, other slower mechanisms of crack growth which operate at much lower stresses or crack sizes. These are discussed in later chapters.

Program 2.4 CRIT: Critical crack size or stress

```
100 PRINT 'COMPUTES CRITICAL STRESS OR CRACK LENGTH FOR'
110 PRINT 'UNSTABLE CRACK GROWTH IF UNKNOWN IS INPUT AS ZERO'
120 PRINT 'TERMINATES IF LENGTH A IS INPUT AS NEGATIVE'
130 REM P1=PI, E=YOUNGS MODULUS (MPA), N=POISSONS RATIO
140 REM M=SURFACE ENERGY DENSITY, GAMMA (MN/M)
150 P1=4.0*ATN(1.0)
160 E=200000
170 N=0.3
180 M=0.012
190 T$="PLANE STRESS"
200 INPUT " PLANE STRESS (Y/N)",A$
210 IF(A$="Y") GOTO 240
220 T$="PLANE STRAIN"
230 E=E/(1-N*N)
240 PRINT
250 INPUT " CRACK LENGTH A (M), APPLIED STRESS (MPA)",A,S
260 IF (A<0) GOTO 360
280 PRINT T$
290 IF (A=0) GOTO 330
300 S=SQRT(2*E*M/(P1*A))
310 PRINT "CRACK LENGTH =";A;"M CRITICAL STRESS =";S;"MPA"
320 GOTO 350
330 A=2*E*M/(P1*S*S)
340 PRINT "STRESS =";S;"MPA CRITICAL CRACK LENGTH =";A;"M"
350 GOTO 240
360 END
```

Sample run 1

```
RUN
CRIT

COMPUTES CRITICAL STRESS OR CRACK LENGTH FOR
UNSTABLE CRACK GROWTH IF UNKNOWN IS INPUT AS ZERO
TERMINATES IF LENGTH A IS INPUT AS NEGATIVE
 PLANE STRESS (Y/N)      ?Y

 CRACK LENGTH A (M), APPLIED STRESS (MPA) ? 0,500
PLANE STRESS
 STRESS = 500  MPA CRITICAL CRACK LENGTH = .611155E-02 M
```

```
 CRACK LENGTH A (M), APPLIED STRESS (MPA)   ? .005,0
PLANE STRESS
 CRACK LENGTH = .005 M CRITICAL STRESS = 552.791  MPA

 CRACK LENGTH A (M), APPLIED STRESS (MPA)   ? -1,0

Ready
```

Sample run 2

```
RUN
CRIT

COMPUTES CRITICAL STRESS OR CRACK LENGTH FOR
UNSTABLE CRACK GROWTH IF UNKNOWN IS INPUT AS ZERO
TERMINATES IF LENGTH A IS INPUT AS NEGATIVE
 PLANE STRESS (Y/N)        ?N

 CRACK LENGTH A (M), APPLIED STRESS (MPA)   ? 0,500
PLANE STRAIN
 STRESS = 500  MPA CRITICAL CRACK LENGTH = .671599E-02 M

 CRACK LENGTH A (M), APPLIED STRESS (MPA)   ? .005,0
PLANE STRAIN
 CRACK LENGTH = .005 M CRITICAL STRESS = 579.482  MPA

 CRACK LENGTH A (M), APPLIED STRESS (MPA)   ? -1,0

Ready
```

Program notes

(1) The critical crack size or stress is evaluated in lines 330 and 300 using equation (2.21)

(2) Material constants have been included in the program for simplicity.

(3) For plane strain $E = E/(1 - \nu^2)$ (line 230). Note that the assumption of plane strain allows a slightly larger crack or stress before failure.

2.5 Crack stresses and displacements

Simply pointing out that a crack tip gives rise to theoretically infinite stresses does not provide a basis for comparing the severity of different crack configurations; some further consideration is required. Complex variable techniques may be used to analyse the behaviour of stresses near a traction-free crack tip in an infinite plate - figure 2.9. This process is discussed by Sih and Liebowitz (1968). Stresses are given

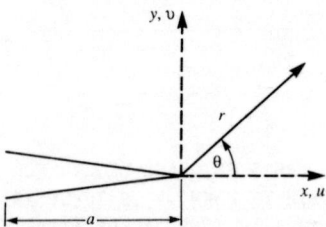

Figure 2.9 Crack tip co-ordinates

in the form of an infinite series, for example:

$$\sigma_y = \frac{K_I}{\sqrt{2\pi r}} f(\theta) + A + O(\sqrt{r}) \qquad (2.22)$$

where A is a constant and $O(\sqrt{r})$ means terms of order $\sqrt{r}$ and higher powers of r. The parameter K_I is the *stress intensity factor* which is a constant for fixed geometry and loading; It is defined in this case as

$$K_I = \lim_{r \to 0} \sqrt{2\pi r} \; \sigma_y \qquad (2.23)$$

Now the first term in the series (2.22) depends on $1/\sqrt{r}$, where r is the distance from the crack tip. Hence when r is small (i.e. close to the crack tip), the first term is much larger than subsequent terms which are therefore usually neglected. The significance of these neglected terms is examined more closely in chapter 5. Note that when r tends to zero the stresses tend to infinity. In real materials, this infinite crack tip stress is never actually attained but the concept of the stress intensity factor is still valuable in the analysis of crack problems.

We now give the first term of the series solution for the stresses and displacements in full. For simplicity the possible movements of the crack surfaces are divided into three distinct cases or modes, which

Crack stresses and displacements

are illustrated in the next three figures. *Mode I* is the opening mode, *mode II* the sliding (or shear) mode and *mode III* the tearing mode.

The expressions are:

Mode I

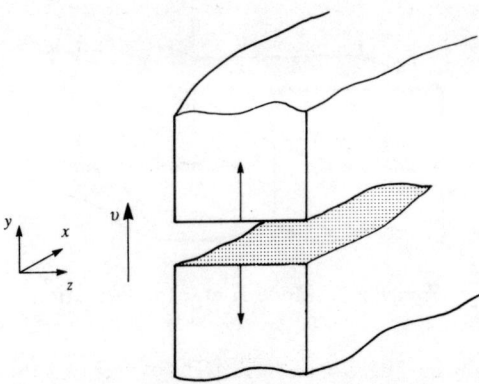

Figure 2.10 Mode I crack deformation

$$\sigma_x = \frac{K_I}{\sqrt{2\pi r}} \cos\frac{\theta}{2} \left(1 - \sin\frac{\theta}{2} \sin\frac{3\theta}{2}\right)$$

$$\sigma_y = \frac{K_I}{\sqrt{2\pi r}} \cos\frac{\theta}{2} \left(1 + \sin\frac{\theta}{2} \sin\frac{3\theta}{2}\right)$$

$$\tau_{xy} = \frac{K_I}{\sqrt{2\pi r}} \sin\frac{\theta}{2} \cos\frac{\theta}{2} \cos\frac{3\theta}{2} \quad (2.24)$$

$$u = \frac{K_I}{4\mu} \sqrt{\frac{r}{2\pi}} \left((2\kappa - 1)\cos\frac{\theta}{2} - \cos\frac{3\theta}{2}\right)$$

$$v = \frac{K_I}{4\mu} \sqrt{\frac{r}{2\pi}} \left((2\kappa + 1)\sin\frac{\theta}{2} - \sin\frac{3\theta}{2}\right)$$

where

$$K_I = \lim_{r \to 0} \sqrt{2\pi r}\ \sigma_y$$

Mode II

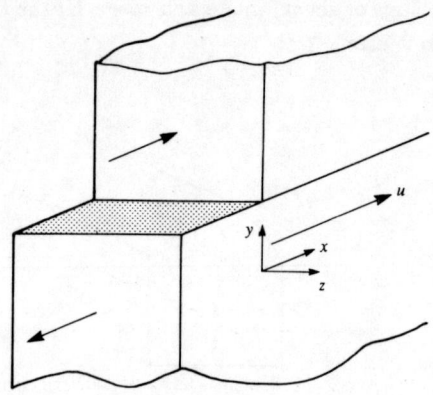

Figure 2.11 Mode II crack deformation

$$\sigma_x = -\frac{K_{II}}{\sqrt{2\pi r}} \sin\frac{\theta}{2} \left(2 + \cos\frac{\theta}{2}\cos\frac{3\theta}{2}\right)$$

$$\sigma_y = \frac{K_{II}}{\sqrt{2\pi r}} \sin\frac{\theta}{2} \left(\cos\frac{\theta}{2}\cos\frac{3\theta}{2}\right)$$

$$\tau_{xy} = \frac{K_{II}}{\sqrt{2\pi r}} \cos\frac{\theta}{2} \left(1 - \sin\frac{\theta}{2}\sin\frac{3\theta}{2}\right) \qquad (2.25)$$

$$u = \frac{K_{II}}{4\mu}\sqrt{\frac{r}{2\pi}} \left((2\kappa + 3)\sin\frac{\theta}{2} + \sin\frac{3\theta}{2}\right)$$

$$v = -\frac{K_{II}}{4\mu}\sqrt{\frac{r}{2\pi}} \left((2\kappa - 3)\cos\frac{\theta}{2} + \cos\frac{3\theta}{2}\right)$$

where

$$K_{II} = \lim_{r \to 0} \sqrt{2\pi r} \ \tau_{xy}$$

Mode III

$$\sigma_x = \sigma_y = \tau_{xy} = 0; \qquad u = v = 0$$

$$\tau_{xz} = -\frac{K_{III}}{\sqrt{2\pi r}} \sin\frac{\theta}{2}$$

Crack stresses and displacements

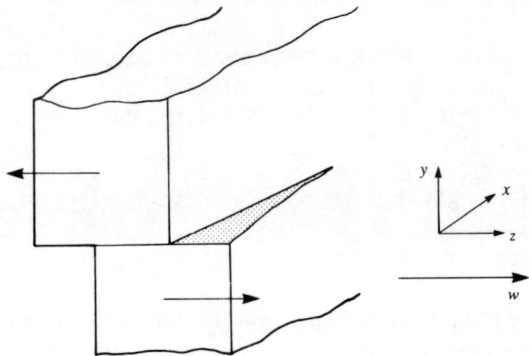

Figure 2.12 Mode III crack deformation

$$\tau_{yz} = \frac{K_{III}}{\sqrt{2\pi r}} \cos\frac{\theta}{2} \qquad (2.26)$$

$$w = 2\frac{K_{III}}{\mu}\sqrt{\frac{r}{2\pi}} \sin\frac{\theta}{2}$$

where

$$K_{III} = \lim_{r \to 0} \sqrt{2\pi r}\ \tau_{zy}$$

In the above equations, the first term of each expansion for stress and displacement is given separately for each mode. If two or three modes are present simultaneously (a *mixed-mode* problem), the corresponding stresses and displacements for each mode may be simply added together which is the *principle of superposition*.

Two-dimensional approximations are commonly used as outlined in section (2.1). For plane strain (which approximates a thick sheet) the formulae for mode I and mode II are used with $\kappa = 3 - 4\nu$. For plane stress, (which approximates a very thin sheet) the same formulae are used with $\kappa = (3 - \nu)/(1 + \nu)$.

The above equations may now be used in a BASIC program to determine stresses and displacements close to the tip of a perfect crack. Only the expressions for mode I and mode II are included allowing a choice of plane stress or plane strain conditions. The program assumes that material properties will not need to be changed frequently and so includes these as program lines 140 and 150. Information about the stress state, the stress intensity factors and the local (r, θ) co-ordinates is entered interactively via INPUT statements.

Program 2.5 CRSS - crack displacements and stresses

```
100 PRINT "DISPLACEMENTS AND STRESSES AT DISTANCE (R),"
110 PRINT "ANGLE (THETA) FROM A CRACK TIP"
120 PRINT "TERMINATES IF (R) IS NEGATIVE. UNITS MN, METRES"
130 PRINT
140 INPUT "  KI     KII ",K1,K2
150 REM P1=PI, E=YOUNGS MODULUS, N=POISSONS RATIO (NU)
160 P1=4.0*ATN(1.0)
170 E=200000
180 N=0.3
190 REM M=SHEAR MODULUS (MU), K=KAPPA FOR PLANE STRESS
200 M=E/(2*(1+N))
210 K=(3-N)/(1+N)
220 T$="PLANE STRESS"
230 INPUT " PLANE STRESS (Y/N)",A$
240 IF(A$="Y") GOTO 280
250 T$="PLANE STRAIN"
260 REM  KAPPA FOR PLANE STRAIN
270 K=3-4*N
280 INPUT " DISTANCE R, ANGLE THETA (DEGREES)";R;A
290 IF (R<=0.0) GOTO 490
300 REM  CONVERT ANGLE TO RADIANS
310 T=A*P1/180
320 C=COS(T/2.0)
330 S=SIN(T/2.0)
340 C3=COS(3.0*T/2.0)
350 S3=SIN(3.0*T/2.0)
360 REM ********** DISPLACEMENTS
370 A1=SQRT(R/(2*P1))/(4*M)
380 U1=A1*(K1*((2*K-1.0)*C-C3)+K2*((2*K+3.0)*S+S3))
390 U2=A1*(K1*((2*K+1.0)*S-S3)-K2*((2*K-3)*C+C3))
400 PRINT T$;"          UX =";U1," UY =";U2
410 REM ********** STRESSES
420 A2=SQRT(2*P1*R)
430 T1=(K1*C*(1.0-S*S3)-K2*S*(2.0+C*C3))/A2
440 T2=(K1*C*(1.0+S*S3)+K2*S*C*C3)/A2
450 T3=(K1*S*C*C3+K2*C*(1.0-S*S3))/A2
460 PRINT " SX =";T1;"     SY =";T2;"     TXY =";T3
470 PRINT
480 GOTO 280
490 END
```

Sample run 1

```
RUN
CRSS

DISPLACEMENTS AND STRESSES AT DISTANCE (R),
ANGLE (THETA) FROM A CRACK TIP
TERMINATES IF (R) IS NEGATIVE.

 KI    KII    ? 100,0
PLANE STRESS (Y/N)        ? Y
DISTANCE R, ANGLE THETA (DEGREES) ? .0001,180
PLANE STRESS         UX =-.194204E-12   UY = .797885E-05
 SX =-.348766E-03    SY = 0      TXY =-.20795E-11

DISTANCE R, ANGLE THETA (DEGREES) ? .0001,0
PLANE STRESS         UX =.27926E-05   UY = 0
 SX =3989.42    SY = 3989.42     TXY = 0

DISTANCE R, ANGLE THETA (DEGREES) ? -1,0
Ready
```

Sample run 2

```
RUN
CRSS

DISPLACEMENTS AND STRESSES AT DISTANCE (R),
ANGLE (THETA) FROM A CRACK TIP
TERMINATES IF (R) IS NEGATIVE.

 KI    KII    ? 100,0
PLANE STRESS (Y/N)        ? N
DISTANCE R, ANGLE THETA (DEGREES) ? .0001,180
PLANE STRAIN       UX =-.162815E-12   UY = .726075E-05
 SX =-.348766E-03    SY = 0     TXY =-.20795E-11

DISTANCE R, ANGLE THETA (DEGREES) ? .0001,0
PLANE STRAIN       UX = .20745E-05   UY = 0
 SX =3989.42     SY = 3989.42     TXY = 0

DISTANCE R, ANGLE THETA (DEGREES) ? -1,0
Ready
```

Sample run 3

```
RUN
CRSS

DISPLACEMENTS AND STRESSES AT DISTANCE (R),
ANGLE (THETA) FROM A CRACK TIP
TERMINATES IF (R) IS NEGATIVE.

 KI    KII   ? 100,50
PLANE STRESS (Y/N)          ? Y
DISTANCE R, ANGLE THETA (DEGREES) ? .0001,180
PLANE STRESS          UX =.398942E-05    UY = .797885E-05
 SX =-3989.42     SY = -.103975E-11    TXY = -.174383E-03

DISTANCE R, ANGLE THETA (DEGREES) ? .0001,0
PLANE STRESS          UX =.27926E-05     UY = -.13963E-05
 SX =3989.42      SY = 3989.42      TXY = 1994.71

 distance R, angle THETA (DEGREES)           ? -1,0
Ready
```

Program notes

(1) When θ is 180 degrees, parameters are being calculated on the crack surface. There should be zero stress on this free surface since no external tractions are applied. The relatively small stresses output by CRSS are errors due to the limited number of decimal places used in BASIC.

(2) Under symmetrical mode I loading, the continuation of the crack line extending through the material is a line of symmetry; There is therefore no shear stress. The formulae also predict that $\sigma_x = \sigma_y$ on this line, and clearly this will only be a good approximation close to the crack tip.

(3) The sample run 2 is simply a repeat of run 1 but assuming plane strain conditions. For plane stress or plane strain the predicted stresses are identical; the displacements are smaller for plane strain which approximates a very thick plate.

(4) In sample run 3, K_{II} is non-zero. This introduces a non-zero σ_x on the crack surface and a non-zero shear stress τ_{xy} within the material.

It is worth re-emphasising that the use of this one term approximation to an infinite series is only valid if r is small. In fact, the error in using only the first term of the series is more than 1 % for $r \geq 0.02a$. We would also point out that the expansions are derived assuming that the elastic material lies between the lines $\theta = \pm 180°$ and therefore only angles in this range should be used.

2.6 References

Iremonger, M.J. (1982) BASIC stress analysis, Butterworths, London
Parker, A.P. (1981) The Mechanics of Fracture and Fatigue, E & F Spon Ltd, London.
Peterson, R.E. (1974) Stress Concentration Factors, John Wiley, London
Sih, G.C. and Liebowitz H. (1968) Mathematical theories of brittle fracture in *Fracture - An Advanced Treatise*, Vol II, Ed. H. Liebowitz, Academic press, New York.

PROBLEMS

(2.1) A thin vertical steel sheet ($E = 205,000 MPa, \nu = 0.3, \gamma = 11 kJ/m^2$) must support a stress of 2,100 Pa on its lower edge. What is the largest crack allowable if the sheet is not to fracture as soon as the load is applied ?

(2.2) A steel sheet of the same material as in 2.1, has a crack of length 0.2 mm at right angles to the direction of loading. What is the least stress that will cause the sheet to fracture immediately assuming (a) plane stress (b) plane strain ?

(2.3) Assuming a sheet as in 2.2 above, what is the effect on the fracture stress of a 10 per cent error in ν ?

(2.4) What is the effect of changing
(a) Young's modulus
(b) Poisson's ratio
on the stresses and displacements near a crack tip ? Consider both plane stress and plane strain approximations.

(2.5) Consider the example of a crack under mode II loading only with $K_{II} = 50 MPa m^{1/2}$. Relate the stresses and displacements obtained to those in sample run 3 of CRSS.

(2.6) Modify program CRSS to input K_{III} and output the resulting stresses and displacements.

(2.7) Use program CRSS to find the distance r at which σ_x has some fixed value by trial and error. Can you alter the program to find this distance directly? Find such a distance for a range of angles and plot the contour of constant stress around the crack tip.

Chapter 3
Stress intensity factors

3.1 Introduction

The stress intensity factor K is a measure of the severity of the defect in a given cracked region and its determination is normally the first step in the process of deciding the maximum allowable crack size or stress. Values of stress intensity factor (SIF) are often normalised with a divisor K_0 ($= \sigma\sqrt{\pi a}$) which is the K-value corresponding to a crack of half-length a in an infinite sheet under a constant normal load σ. The ratio K_I/K_0 is referred to as the *configuration correction factor* Q, since it is a measure of how the geometry and loading of a particular configuration affect the SIF for a crack in an infinite plate. Using this nomenclature we have:

$$K_I = Q\sigma\sqrt{\pi a} \qquad (3.1)$$

For an arbitrary region it will usually be impossible to calculate the SIF analytically. Numerical methods can be used but they may be expensive in terms of computer time. If existing data can be applied to a given problem it will usually be the fastest and cheapest approach.

Note that we do not consider problems where the crack surfaces would meet under load. This type of problem, known as *crack closure*, must be treated as a special case and may result in a negative value of K_I.

3.2 Using tables and graphs

Many practical examples of crack problems are similar in geometry and loading to a standard configuration for which stress intensity factors are already available. If this is the case, we can obtain the SIF for the standard configuration and adapt it to the particular example. Tables and graphs of SIFs for a range of problems have been printed in bookform such as Tada et al (1973) or Rooke and Cartwright (1976) and should be available in a good engineering library. A typical graph is shown in figure 3.1 for the case of a crack in a rectangular sheet with constant normal loading on the ends of the sheet. Curves for several different relative sheet dimensions are shown and the Q-value for these dimensions is simply read from the graph.

Suppose we have a square sheet of side $1m$ with a crack of length $0.4m$ and a stress of $10Pa$ on the ends of the sheet; then $h/b = 1$ and $a/b = 0.4$. Reading from the curve for $h/b = 1$ gives $K_I/K_0 = 1.21$. To get the actual stress intensity factor for these particular dimensions multiply by K_0 giving $K_I = 7.93 Pa m^{1/2}$.

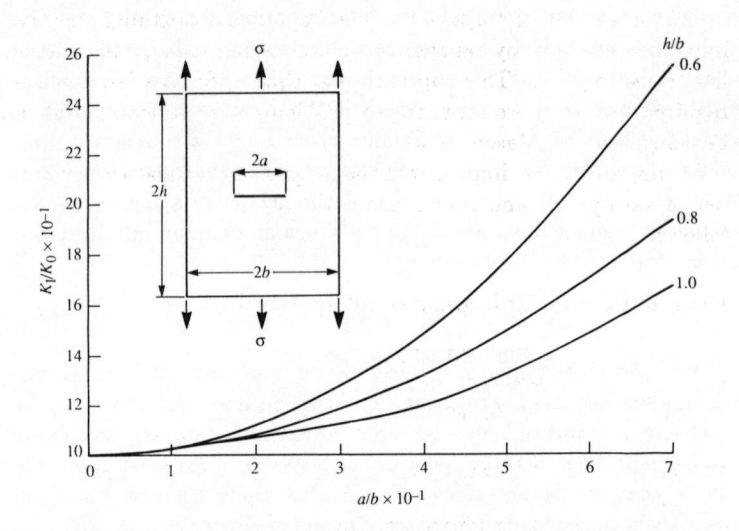

Figure 3.1 K for a central crack

Stress intensity factors for intermediate dimensions must be calculated by *interpolation*. The simplest approach being *Lagrange interpolation* in which a polynomial is found which goes through a set of given points. If two points are given, the equation of a straight line which passes through both points can be determined. Intermediate values are then found from this equation. Similarly, with three points, a quadratic curve may be fitted and so on. The Lagrange interpolation formula assumes there are n data points (x_i, y_i), $i = 1, n$ and gives the polynomial passing through these points as:

$$p(x) = \sum_{i=1}^{n} y_i \, l_i(x) \qquad (3.2)$$

where

$$l_i(x) = \prod_{\substack{j=1 \\ j \neq i}}^{n} \frac{(x - x_i)}{(x_i - x_j)}$$

$l_i(x)$ is simply the product of all terms like $(x - x_i)/(x_i - x_j)$ with the brackets $(x - x_i)/(x_i - x_i)$ omitted. Once an equation in the form (3.2) has been determined then values at any value of x can be calculated. It will also be useful to have an explicit expression for $p(x)$ in the form $a_0 + a_1 x + a_2 x^2 + \ldots$. The Lagrange algorithm is therefore implemented below by expanding the brackets in each l_i and collecting like terms together. This approach is adequate if just a few terms are required but for more complicated problems a numerical text should be consulted (e.g Mason 1983).

A program which implements the Lagrange formula for any number of data points and allows calculation of intermediate values is as follows.

Program 3.2 LAG: Lagrangian interpolation

```
100 PRINT " LAGRANGE INTERPOLATION"
110 DIMENSION X(6),Y(6),C(6),T(6)
120 INPUT ' NUMBER OF POINTS TO BE GIVEN';N
130 FOR I=1 TO N
140 INPUT ' XI,YI' ;X(I),Y(I)
150 NEXT I
160 REM *** FIND COEFFICIENTS OF POLYNOMIAL
170 FOR I=1 TO N
180 Q=1
190 M=0
200 FOR K=1 TO N
210 C(K)=1
220 NEXT K
230 FOR J=1 TO N
240 IF J=I THEN 320
250 IF M=0 THEN 290
260 FOR K=M TO 1 STEP -1
270 C(K+1)=-X(J)*C(K+1)+C(K)
```

```
280 NEXT K
290 C(1)=-X(J)*C(1)
300 Q=Q*(X(I)-X(J))
310 M=M+1
320 NEXT J
330 Q=Y(I)/Q
340 FOR K=1 TO N
350 T(K)=T(K)+C(K)*Q
360 NEXT K
370 NEXT I
380 REM *** PRINT AND EVALUATE POLYNOMIAL
390 FOR K=1 TO N
400 PRINT " COEFFICIENT OF X**";(K-1);" IS ";T(K)
410 NEXT K
420 INPUT ' NEW X VALUE';X1
430 Y1=0
440 FOR I=1 TO N
450 Y1=Y1+T(I)*X1**(I-1)
460 NEXT I
470 PRINT ' AT X =';X1;'   Y =';Y1
480 PRINT
490 INPUT ' ANOTHER X-VALUE (Y/N)';A$
500 IF A$='Y' GOTO 420
510 END
```

Sample run 1

```
RUN LAG

 LAGRANGE INTERPOLATION
 NUMBER OF POINTS TO BE GIVEN ? 2
 XI,YI ? 0.8,1.32
 XI,YI ? 1.0,1.21
 COEFFICIENT OF X** 0  IS  1.76
 COEFFICIENT OF X** 1  IS -.55
 NEW X VALUE ? 0.8
 AT X = 0.8    Y = 1.32

 ANOTHER X-VALUE (Y/N)? Y
 NEW X VALUE ? 0.9
 AT X = 0.9    Y = 1.265

 ANOTHER X-VALUE (Y/N)? N
```

Ready

Sample run 2

```
RUN LAG

 LAGRANGE INTERPOLATION
 NUMBER OF POINTS TO BE GIVEN ? 3
 XI,YI ? 0.6,1.48
 XI,YI ? 0.8,1.32
 XI,YI ? 1.0,1.21
 COEFFICIENT OF X** 0   IS  2.26
 COEFFICIENT OF X** 1   IS -1.675
 COEFFICIENT OF X** 2   IS  .625
 NEW X VALUE ? 1.0
 AT X = 1.0     Y = 1.21

 ANOTHER X-VALUE (Y/N)? Y
 NEW X VALUE ? 0.9
 AT X = 0.9     Y = 1.25875

 ANOTHER X-VALUE (Y/N)? N
Ready
```

Program notes

(1) The input data points have been taken from figure 3.1 for the example with $a/b = 0.4$ given above.

(2) The program returns the correct values at the data points as expected, and the linear and quadratic interpolated values for $h/b = 0.9$. The quadratic value should be the most accurate. The accuracy cannot, of course, be greater than the original data on which the graph is based.

(3) Note that values computed far from the original data points may be unreliable. Using more than four or five data points requires careful consideration since the Lagrange polynomial will then have several maximum or minimum points and may then be rather more 'wiggly' than the data would suggest.

3.3 Superposition

The *principle of superposition* applies to any linear elastic system. If two or more different loadings are applied to the system, the effect of the combined loads is the sum of their individual effects. In particular, the stress intensity factor for a region under multiple loading is the sum of the SIFs for the region under each part of the load. The principle is illustrated in figure 3.2. In part (a), a cracked body is subject to

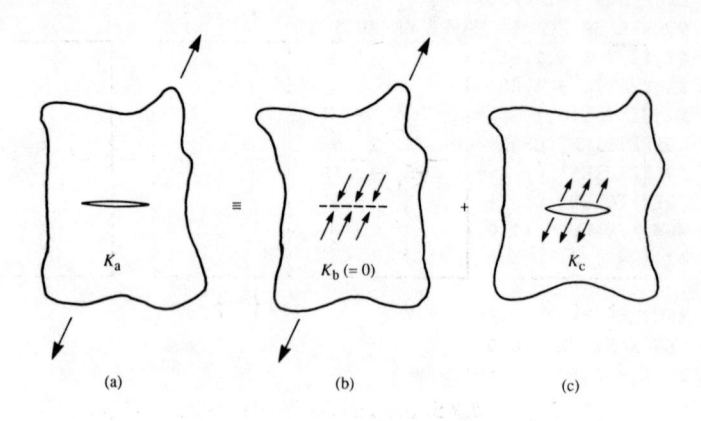

Figure 3.2 The principle of superposition

some load system which causes the crack to open and gives rise to a stress intensity factor K_a. In part (b), the crack is not present but the stresses on the crack site are measured as $-\sigma_x$. Equivalently, we can consider the crack to be present but prevented from opening by a system of stresses $-\sigma_x$. Since the crack does not open, the SIF for this case $K_b = 0$. Finally, in part (c), the stresses closing the crack in (b) are applied in reverse thus opening the crack and giving rise to a stress intensity factor K_c. The principle of superposition can be applied to the stress or displacement at any part of the region giving:

$$\sigma_a = \sigma_b + \sigma_c \tag{3.3}$$

$$u_a = u_b + u_c \tag{3.4}$$

In particular, the stress intensity factor is

$$K_a = K_b + K_c \tag{3.5}$$

Superposition

The stress intensity factor for any linear elastic region may therefore always be found from the original problem or from the same geometry with no external loads and the appropriate stresses applied to the crack surface. If the SIF is to be obtained from a number of different cracked configurations, then any of these may have overlapping crack surfaces providing that the total configuration does not.

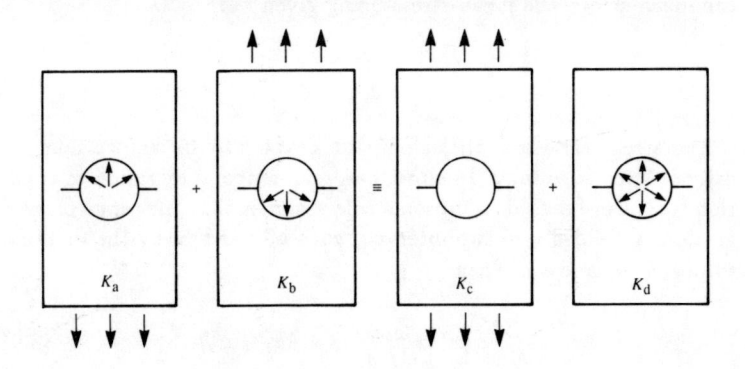

Figure 3.3 Superposition for a pin-loaded hole

Another example of superposition is shown in figure 3.3. This example allows us to calculate the SIF for a plate under tension in one direction supported by a pin in a cracked hole from the two symmetric cases of constant tension on the ends of the plate and constant pressure on the hole (or vice versa). The relationship between the SIFs is

$$K_a = K_b = \frac{1}{2}(K_c + K_d) \qquad (3.6)$$

3.3.1 Approximations using superposition

The principle of superposition enables us to relate K-values for different load cases when the geometries are the same. If the geometries are different then the principle does not strictly apply but may be used to obtain approximate solutions. A common example is that of a crack at the edge of a cutout in a plate. If the crack is small we can approximate the geometry (by neglecting the curvature near the crack) and the loading (by assuming that it is constant). The SIF may now be estimated by using the known SIF for an edge crack in an infinite sheet under constant normal pressure σ, that is

$$K_I = 1.12\sigma\sqrt{\pi a} \tag{3.7}$$

where a is the length of the crack. The stress σ may be estimated using the maximum stress on the crack, the stress on the crack tip or the mean stress, the mean stress being given by:

$$\sigma_m = \frac{1}{a} \int_0^a \sigma(x) dx \tag{3.8}$$

The stress variation $\sigma(x)$ along the crack may be known analytically or experimentally: In either case, numerical integration of equation (3.8) can be used. Simpson's rule assumes that the range of integration is divided into two intervals, each of width h, by the function values σ_0, σ_1 and σ_2. Then:

$$\int_0^{2h} \sigma(x) dx \simeq \frac{h}{3}(\sigma_0 + 4\sigma_1 + \sigma_2) \tag{3.9}$$

If greater accuracy is required then the range of integration is divided into a number of pairs of intervals and the final approximation is computed by summing all these results. This integration method is used in program 3.3 to find the mean stress on the site of a crack emanating from a circular hole of radius R at right angles to an applied stress σ_0 (figure 3.4). The stress in the absence of the crack is given

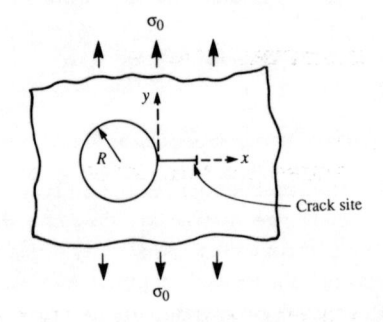

Figure 3.4 Hole in an infinite sheet

by:
$$\sigma(x) = \sigma_0 \left[1 + \frac{1}{2}\left(\frac{R}{R+x}\right)^2 + \frac{3}{2}\left(\frac{R}{R+x}\right)^4\right] \quad (3.10)$$

Note that $x = 0$ is the edge of the hole, and equation (3.10) gives the stress as $3\sigma_0$ which is the correct stress concentration factor in this case. Program 3.3 also computes the crack tip stress and the maximum stress (which is at the edge of the hole). These stresses are then multiplied by 1.12 to give an edge-crack approximation of the Q-value.

Program 3.3 MS: Approximate K-values from mean stress

```
100 PRINT " MS - DETERMINATION OF K FROM MEAN STRESS"
110 INPUT "   RADIUS OF HOLE R";R
120 INPUT "   LENGTH OF CRACK A ";A
130 PRINT
140 INPUT "  EVEN NUMBER OF INTERVALS (ZERO TO STOP)";N
150 IF N=0 GOTO 340
160 H=A/N
170 DEF FNS(X)=1+0.5*(R/(R+X))**2+1.5*(R/(R+X))**4
180 REM *** FIND INTEGRAL OF S(X) USING SIMPSONS RULE
190 C=0
200 S2=FNS(X0)
210 FOR I=0 TO (N-2) STEP 2
220 S0=S2
230 S1=FNS(X0+(I+1)*H)
240 S2=FNS(X0+(I+2)*H)
250 C=C+S0+4*S1+S2
260 NEXT I
270 C=C*H/3
280 REM *** FIND STRESSES AND SIFS
290 M=C/A
300 L=FNS(X0)
310 T=FNS(X0+N*H)
320 PRINT "MEAN STRESS=";M;" MAX STRESS=";L;" TIP STRESS=";T
330 GOTO 130
340 M=1.12*M
350 L=1.12*L
360 T=1.12*T
370 PRINT " MEAN Q =";M;" MAX Q =";L;" TIP Q =";T
380 END
```

Sample run 1

RUN MS

MS - DETERMINATION OF K FROM MEAN STRESS
  RADIUS OF HOLE R? 1
  LENGTH OF CRACK A ? .01

  EVEN NUMBER OF INTERVALS (ZERO TO STOP)? 4
MEAN STRESS= 2.96554 MAX STRESS= 3 TIP STRESS= 2.93162

  EVEN NUMBER OF INTERVALS (ZERO TO STOP)? 8
MEAN STRESS= 2.96554 MAX STRESS= 3 TIP STRESS= 2.93162

  EVEN NUMBER OF INTERVALS (ZERO TO STOP)? 0
  MEAN Q = 3.32141 MAX Q = 3.36 TIP Q = 3.28341
Ready

Sample run 2

RUN MS

MS - DETERMINATION OF K FROM MEAN STRESS
  RADIUS OF HOLE R? .01
  LENGTH OF CRACK A ? 1

  EVEN NUMBER OF INTERVALS (ZERO TO STOP)? 50
MEAN STRESS= 1.01645 MAX STRESS= 3 TIP STRESS= 1.00005

  EVEN NUMBER OF INTERVALS (ZERO TO STOP)? 100
MEAN STRESS= 1.01164 MAX STRESS= 3 TIP STRESS= 1.00005

  EVEN NUMBER OF INTERVALS (ZERO TO STOP)? 0
  MEAN Q = 1.13303 MAX Q = 3.36 TIP Q = 1.12005
Ready

Program notes

(1) The stress is determined by a function in line 170. If necessary, stresses could be read in point by point from a graph or table.
(2) The program allows integration to be repeated with an increased number of points. Little change in the mean stress indicates that the

result is satisfactory. A large change suggests that the integration should be repeated again with even more points.

(3) To obtain actual stress intensity factors the Q-values should be multiplied by $K_0 = \sigma\sqrt{\pi a}$. This multiplication could be simply added to the program if desired.

(4) In the first run, the crack is very small compared to the hole and should, therefore, be well approximated by an edge crack in an infinite sheet with an applied stress of three units. In other words, for small edge cracks a good estimate of SIF is simply the stress concentration factor times the edge cracked plate value. Note that very few integration points are required. All three approximations give acceptable values.

(5) In the second run, the hole is very small compared to the crack, there is little stress concentration and the result is very close to that for an edge crack in an infinite sheet. However, the function for stress is now more difficult to integrate and many more points are required. The maximum stress approximation gives a rather poor approximation to the stress intensity factor.

It is important to realise that values obtained using the methods in this section are approximations. The effect of interacting boundaries may be highly significant. For example, if the hole in the above problem is in a plate four times the width of the hole, then the stress concentration increases from 3 to 3.23 with a corresponding change in SIF. If the plate is only twice the width of the hole, the stress increases to 4.32 times the applied stress. Before tackling 'real' problems it is necessary to build up some experience solving problems with known solutions to discover what constitutes a reasonable approximation.

3.4 Numerical methods

The analytical solution of elasticity problems is only possible in a few simple cases. To obtain a solution to realistic problems some form of approximation of the actual problem is necessary - a numerical method. There are many such methods for the solution of elastic fracture problems but they can rather crudely be classified as general or specialised methods.

A general method should be capable of modelling an arbitrary two- or three-dimensional elastic region containing one or more cracks or holes of any shape or size. A specialised method may require a new mathematical form to cater for a different crack or component shape.

Specialised methods are of immense value to the study of fracture problems since they tend to be more accurate and more efficient than general methods. They provide 'benchmark' solutions of high accuracy against which other methods can be judged. The assessment of accuracy is difficult since the exact solution is usually unknown, but errors are often estimated by increasing the number of parameters used in the approximation. Some references are given by Parker (1981).

The collocation method may only be applied to two-dimensional problems and approximates the solution as a complex series expansion; as more terms are included in the expansion accuracy increases and the changes enable errors in the solution to be estimated.

The specialised integral equation method appears in a variety of forms, usually for straight or plane cracks in infinite regions. Accuracy may be increased by increasing the number of points used for numerical integration.

General numerical methods are commonly used for the solution of elasticity and fracture problems in two forms - the finite element method (FEM) and the boundary element (or boundary integral equation) method (BEM). These methods are widely available in the form of computer packages which solve a wide range of two- and three-dimensional problems whilst insulating the user from the mathematical and numerical formulation. Since these packages typically involve tens of man years of work, it is impractical for the novice to consider writing his own program. The formulation of these methods will not therefore, be discussed here. However, we can consider briefly how elastic regions and cracks are modelled since this is vital to the accuracy of results.

3.4.1 The boundary element method

In section 3.1 a curve is approximated (or modelled) by a polynomial which passes through a number of sample points on the true curve. Since polynomials are easy to deal with mathematically, they may be used to approximate an integral form of the rather complicated equations of elasticity. The simplified equations can then be solved numerically. This is the basis of the finite element (FEM) and boundary element (BEM) methods.

The boundary element method uses a different formulation of the elasticity equations to the finite element method, which requires the evaluation of more complicated integrals - but only over the bound-

Numerical methods

ary or surface of the elastic region. Only the surface geometry and boundary conditions need be modelled giving much simpler mesh design and more accurate solutions. The first step in producing a BEM solution is to divide up the boundary into a *mesh* of small segments or *elements*.

As an example, consider a (two-dimensional) rectangular plate under tension which contains a circular hole. Using the problem symmetry this is modelled with a mesh of straight line (linear) elements in figure 3.5. On each element the value of a variable (the displacement u

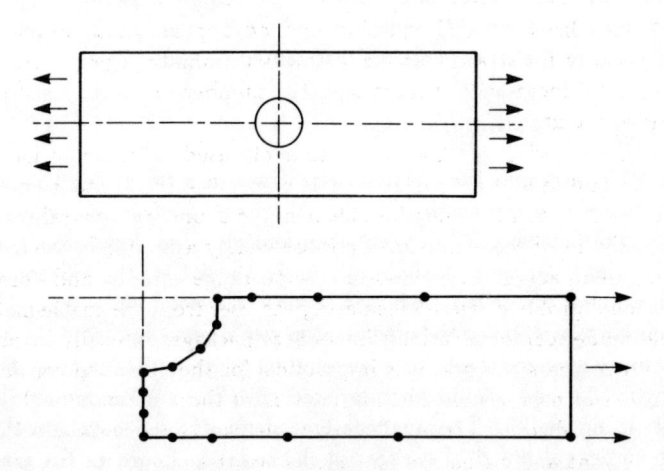

Figure 3.5 A circular hole problem and the corresponding quarter-plate boundary element mesh

say) may be defined by the value at the endpoints (the *nodes* of the element) and a distance along the element denoted by ξ. So using Lagrange interpolation and assuming that ξ varies from 0 to 1 on each element as we go anticlockwise around the boundary we have

$$u = u_1\phi_1(\xi) + u_2\phi_2(\xi) \tag{3.11}$$

where $\phi_1(\xi) = (1 - \xi)$, and $\phi_2(\xi) = \xi$. Note that when $\xi = 0$, $u = u_1$ and when $\xi = 1$, $u = u_2$. The other boundary variables are expressed with the same *interpolation functions* ϕ_1 and ϕ_2 on an *isoparametric* element. Providing that values of x, y and either u or the traction t are given on each node, the computer program can evaluate these variables at any point on the element and hence produce approximations to the required integrals.

Once integrals have been computed over each element, they are assembled into a system of linear equations for the unknown boundary values which is usually solved by Gauss elimination. If stresses and displacements are required at interior points, an additional boundary integration must be performed.

The accuracy of solution increases as the number of elements increases but so does the computing cost! The best mesh therefore uses more elements where any of the boundary variables are changing rapidly and fewer elsewhere. Good mesh design requires some intuition about the form of the solution and some experience in developing meshes so that elements can be distributed accordingly.

3.4.2 The finite element method

The FEM arranges the elasticity equations in a form which requires integration over the whole of the elastic region for solution. Elements for a two-dimensional body are therefore also two-dimensional, typically triangular or quadrilateral. As in the BEM, the variables on each element are usually assumed to vary in a polynomial form based on nodal values, but now, of course, in two-dimensions.

A finite element mesh for a rectangular plate with a hole is shown in figure 3.6 and should be compared with the corresponding BEM mesh in figure 3.5. The quadrilateral elements used have a node at each corner (although these are not shown in the figure for the sake of clarity) and a linear variation is assumed on each side of the element. Even for this simple example it is clear that mesh design is much more

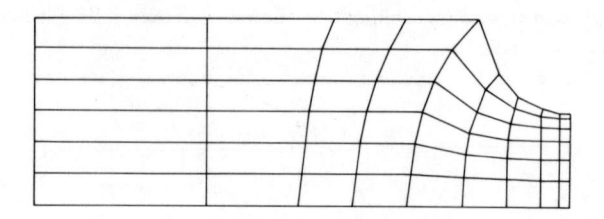

Figure 3.6 A finite element mesh

straightforward for the BEM than for the FEM.

As in the BEM, integration over all elements produces a system of linear equations which are solved for the unknown values on element nodes. The FEM produces a larger system of equations than the BEM but with relatively few non-zero coefficients so that a fast solution method can be used. The solution is obtained at all the nodes, both internal and external, which may make the FEM more efficient than the BEM if the internal solutions are actually needed. However, for the solution of elastic crack problems normally only boundary solutions are required.

If the problem to be analysed includes non-linear (e.g plastic) behaviour, then the FEM can use a similar mesh to the elastic one. The BEM requires additional internal elements to model the non-linear region and is likely to be less efficient than the FEM in such cases.

3.4.3 Finite and boundary elements for crack problems

To use the FEM or BEM for the solution of crack problems requires careful mesh design. The presence of a singularity (the crack tip) drastically reduces the rate at which the numerical solution converges to the exact solution as the number of elements is increased. Many approaches have been used in an attempt to overcome this difficulty and we first consider the use of simple polynomial elements.

Crack modelling using isoparametric elements

Elastic regions with curved boundaries and rapidly changing stresses or displacements are not well-represented by linear elements. By using three nodes per element, parameters can be assumed to vary quadratically and hence model changes as a curve, rather than a straight line. A quadratic boundary element is shown in figure 3.7. The value of

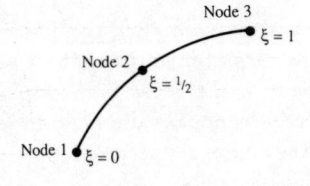

Figure 3.7 A quadratic boundary element

displacement u is again written in terms of nodal values:

$$u = u_1\phi_1 + u_2\phi_2 + u_3\phi_3 \qquad (3.12)$$

where $\phi_1(\xi) = (1-2\xi)(1-\xi)$, $\phi_2(\xi) = 4\xi(1-\xi)$, $\phi_3(\xi) = -\xi(1-2\xi)$. x, y and t are similarly expressed by their nodal values. In the finite element method, each side of a triangular or quadrilateral element will reduce to a similar form to the quadratic boundary element above.

Near the crack tip, we know from chapter 2 that the displacement is proportional to $\sqrt{r}$ and the stress is proportional to $1/\sqrt{r}$. This behaviour is not well represented by polynomial elements and therefore a graded mesh must be used with elements becoming smaller as the crack tip is approached. A large number of elements are required to obtain an accurate value for the stress intensity factor.

Crack modelling using quarter-point elements

Much greater accuracy can be obtained from BEM and FEM models if special elements are introduced at the crack tip which are designed to represent the known $1/\sqrt{r}$ and $\sqrt{r}$ behaviour of stresses and displacements. The most common type of special element is the quarter-point element since these may be introduced without changing the computer program. The middle node of a quadratic element is shifted to the quarter-point position shown in figure 3.8 which distorts the quadratic polynomial to a quadratic in $\sqrt{r}$.

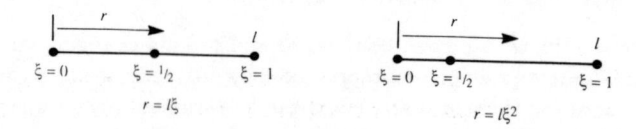

Figure 3.8 Quadratic and quarter-point elements

In the FEM stresses are computed by differentiating element displacements and if the displacement is of the correct form, then so are the stresses. However, in the BEM both displacements and stresses are represented directly by nodal values and to represent the $1/\sqrt{r}$ behaviour of stresses, the computer program must be modified to include such traction-singular elements.

It is very important to note that since the correct crack tip behaviour is only modelled on the special elements on the crack tip, the length

Numerical methods

of these elements is significant. For both FEM and BEM methods a crack tip element length of $0.1a$ to $0.3a$ where a is the half crack length seems to give optimal results without much variation in the computed stress intensity factor.

If we consider the simple problem of a rectangular plate with a central crack shown in figure 3.9 then because of symmetry only a quarter of the plate need be modelled.

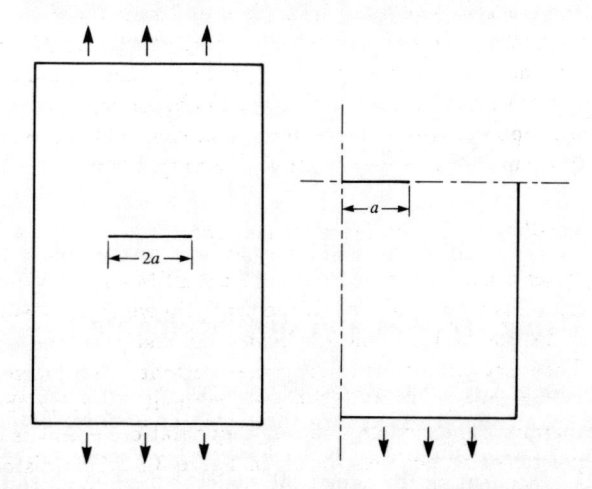

Figure 3.9 A centre-cracked plate and a quarter-plate model

A BEM mesh for the quarter-plate problem is shown in figure 3.10 together with the corresponding FEM mesh. Elements are quadratic in both cases, except at the crack tip where quarter-point elements are used. The FEM should use quarter-point elements on all lines radiating from the crack tip and triangular (collapsed quadrilateral) elements are best. The BEM uses a quarter-point element on the crack surface and a traction-singular quarter-point element joining the crack tip from within the plate material, i.e. on the symmetry line. The error in the stress intensity factor computed from displacements on the above meshes is about 2 percent for the BEM and several times greater for the FEM. Even more accurate BEM variants are available and are reviewed in Smith (1988), FEM methods are reviewed by Gallagher (1978).

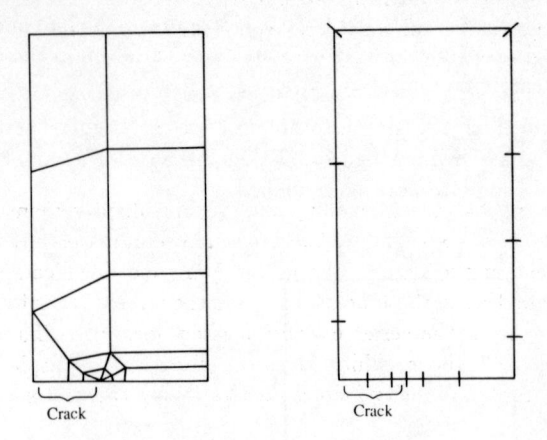

Figure 3.10 BEM and FEM meshes for a quarter-plate model

3.5 Using stresses and displacements

If stresses or displacements can be obtained using a numerical method to model a cracked region then these may be used to compute stress intensity factors directly. The accuracy of numerical methods will be highly dependent on the numerical model or mesh used, and will usually be much less than for the same method without a crack present. In practice, displacements of the crack surfaces are usually the most reliable for SIF calculations. Consider the crack shown in figure 3.11.

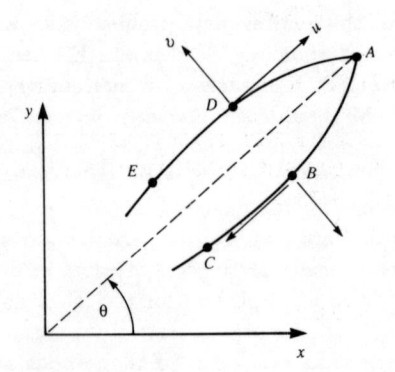

Figure 3.11 Crack tip element nodes

Using stresses and displacements

Assuming the output from our numerical solution is in terms of x and y displacements, we must first calculate the average crack opening displacements, i.e.

$$u_x = (u_{xD} - u_{xB})/2 \tag{3.13}$$

$$u_y = (u_{yD} - u_{yB})/2$$

Calculating the crack opening and sliding displacements using the differences in nodal values automatically eliminates any rigid body movement of the crack. We now require the displacements normal and tangential to the crack tip so that the stress intensity factors can be found from the series expansion given in section 2.5. Referring to figure 3.11 above, which gives the directions of displacements for positive stress intensity factors, and resolving the vectors we obtain:

$$v = u_y \cos\theta - u_x \sin\theta \tag{3.14}$$

$$u = u_y \sin\theta + u_x \cos\theta$$

Taking the angle from the crack tip as 180 degrees in equations (2.13) and (2.14), and rearranging gives:

$$K_I = \frac{2\mu}{\kappa+1}\sqrt{\frac{2\pi}{r}}\ v \tag{3.15}$$

$$K_{II} = \frac{2\mu}{\kappa+1}\sqrt{\frac{2\pi}{r}}\ u$$

A simple program which implements the above calculations for stress intensity factors is as follows:

Program 3.5 KDISP: K_I and K_{II} values from nodal displacements.

```
100 PRINT " KI AND KII FROM CRACK SURFACE DISPLACEMENT"
110 REM P1=PI, E=YOUNGS MODULUS, N=POISSONS RATIO (NU)
120 REM UNITS MN, METRES
130 P1=4.0*ATN(1.0)
140 E=200000
150 N=0.3
160 REM M=SHEAR MODULUS (MU), K=KAPPA FOR PLANE STRESS
170 M=E/(2*(1+N))
180 K=(3-N)/(1+N)
190 T$="PLANE STRESS"
```

```
200 INPUT " PLANE STRESS (Y/N)",A$
210 IF A$="Y" GOTO 250
220 T$="PLANE STRAIN"
230 REM   KAPPA FOR PLANE STRAIN
240 K=3-4*N
250 INPUT " ANGLE OF CRACK TIP TO X-AXIS (DEGREES)";T
260 INPUT " DISTANCE OF POINT FROM CRACK TIP";R
270 INPUT " X DISPLACEMENTS XD,XB ";X1,X2
280 INPUT " Y DISPLACEMENTS YD,YB ";Y1,Y2
290 X=(X1-X2)/2
300 Y=(Y1-Y2)/2
310 C=2*M*SQRT(2*P1/R)/(K+1)
320 REM CHANGE TO LOCAL CRACK CO-ORDINATES
330 T=T*P1/180
340 U=Y*SIN(T) + X*COS(T)
350 V=Y*COS(T) - X*SIN(T)
360 K1=C*V
370 K2=C*U
380 PRINT
390 PRINT T$," KI =";K1;"  KII =";K2
400 END
```

Sample run 1

```
RUN KDISP

 KI AND KII FROM CRACK SURFACE DISPLACEMENT
 PLANE STRESS (Y/N)         ? Y
 ANGLE OF CRACK TIP TO X-AXIS (DEGREES) ? 0
 DISTANCE OF POINT FROM CRACK TIP ? .0001
 X DISPLACEMENTS XD,XB ? 0,0
 Y DISPLACEMENTS YD,YB ? .159577E-4,0

 PLANE STRESS   KI = 100   KII = 0
Ready
```

Using stresses and displacements 53

Sample run 2

```
RUN KDISP

 KI AND KII FROM CRACK SURFACE DISPLACEMENT
 PLANE STRESS (Y/N)          ? Y
 ANGLE OF CRACK TIP TO X-AXIS (DEGREES) ? 45
 DISTANCE OF POINT FROM CRACK TIP ? .0001
 X DISPLACEMENTS XD,XB ? 0,0
 Y DISPLACEMENTS YD,YB ? .159577E-4,0

 PLANE STRESS  KI = 70.7107  KII = 70.7107
Ready
```

Sample run 3

```
RUN KDISP

 KI AND KII FROM CRACK SURFACE DISPLACEMENT
 PLANE STRESS (Y/N)          ? Y
 ANGLE OF CRACK TIP TO X-AXIS (DEGREES) ? 0
 DISTANCE OF POINT FROM CRACK TIP ? .0001
 X DISPLACEMENTS XD,XB ? .398942E-5,-.398942E-5
 Y DISPLACEMENTS YD,YB ? .797885E-5,-.797885E-5

 PLANE STRESS  KI = 100  KII = 50
Ready
```

Program notes

(1) The first part of this program which selects constants appropriate to plane stress or plane strain is identical to that in previous programs.
(2) In the sample runs, displacements have been chosen so that results may be compared with program CRSS in section 2.5.
(3) In run 2, the inclined crack under unidirectional loading gives rise to a mixed-mode problem, that is both K_I and K_{II} are non-zero. Note that for a crack inclined at 45 degrees to the x-axis which is only opening and not sliding, the y-displacement will be positive and the x-displacement negative on the upper surface.
(4) In run 3, a mixed-mode loading gives non-zero values for both K_I and K_{II}.

3.5.1 Displacement extrapolation

The stress intensity factor is defined by an expression of the form

$$K = \lim_{r \to 0} \sigma\sqrt{2\pi r}$$

The displacement extrapolation technique attempts to obtain this limiting value by plotting a series of K-values computed at different distances r from the crack tip. A program such as KDISP above could be used to obtain K-values from displacements on element nodes. A typical plot is shown in figure 3.12. The plot is then extrapolated

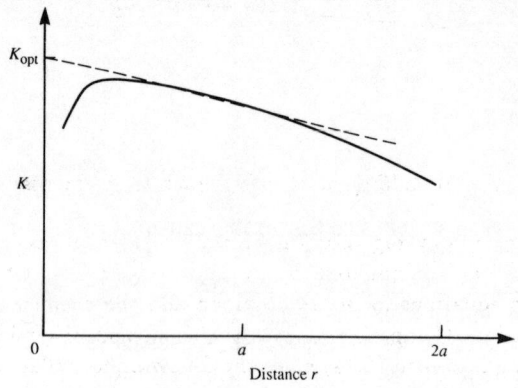

Figure 3.12 Displacement extrapolation

back to $r = 0$ to obtain the 'optimal' value of K. The disadvantage of this approach is that a suitable series of displacement values must be found sufficiently close to the crack tip that r is 'small' and yet not so close that numerical accuracy is lost. This means that a number of elements must be placed close to the crack tip increasing the computer time required to obtain a solution.

3.5.2 Quarter-point elements

If a finite or boundary element method using quarter-point elements has been used to obtain crack displacements, then the previous program KDISP may be used to obtain values of K from the displacements at the quarter-point nodes. Alternatively, the coefficient of $\sqrt{r}$ may be estimated from the element displacements and then used to compute K.

In the previous section the displacements corresponding to a quarter-point element were given in the form:

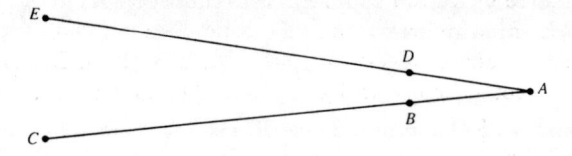

$$u = a_0 + a_1\sqrt{r} + a_2 r \tag{3.16}$$

Figure 3.13 Quarter-point crack tip elements

By finding the average crack opening displacement as in equation the constant a_0 is eliminated. From the average COD at two pairs of nodes, two equations for u are obtained and the coefficient a_1 found by elimination. Comparing a_1 with the coefficient of $\sqrt{r}$ in the series expansions (2.13) and (2.14) gives the following formulae for K:

$$K_I = 2\frac{\mu}{\kappa+1}\sqrt{\frac{2\pi}{l}}(4v_{DB} - v_{EC}) \tag{3.17}$$

$$K_{II} = 2\frac{\mu}{\kappa+1}\sqrt{\frac{2\pi}{l}}(4u_{DB} - u_{EC})$$

Although this approach may seem theoretically sound since we actually find the coefficient of $\sqrt{r}$ rather than simply assuming that r is small and neglecting the r term, in practice the single node formula in KDISP works rather well for quarter-point elements. Note that the quarter-point form is strictly only valid on straight elements although it has also been used successfully on curved elements.

3.6 Energy Relationships

In chapter 2 the condition for sudden crack growth was explained in terms of an energy balance involving the strain energy release rate G, given for plane stress as:

$$G = \frac{\partial U}{\partial a} = \frac{\pi}{E}\sigma^2 a \qquad (3.18)$$

whilst for plane strain E becomes $E/(1-\nu^2)$.

Irwin showed that a relationship exists between G and K_I. A simple, though hardly rigorous approach, is to substitute $K_I = \sigma\sqrt{(\pi a)}$ in the above equation giving

$$G_I = \frac{K_I^2}{E} \qquad (3.19)$$

For pure mode II and pure mode III loading there are similar expressions:

$$G_{II} = \frac{K_{II}^2}{E}$$

$$G_{III} = (1+\nu)\frac{K_{III}^2}{E} \qquad (3.20)$$

It is important to appreciate that the above expressions relating G and K are only applicable to single-mode loading where a straight-fronted crack extends in a plane normal to the load. More complex situations are discussed in chapter 5. To apply the above formulae to determine stress intensity factors, the energy method is normally used in conjunction with numerical techniques such as the boundary element or finite element methods. The cracked region is modelled in the usual way, but the same region is analysed twice, firstly with a crack length a and secondly, with a crack length $a + \delta a$ where δa is a small increment of crack length. The total energy of the region for each analysis (U_1 and U_2) is computed from the displacements and stresses over the whole region. Then:

$$G = \frac{\partial U}{\partial a} = \frac{U_1 - U_2}{\delta a} \qquad (3.21)$$

and hence K may be determined. The main advantage of this method is that the use of special crack tip elements is not essential. Note, however, that two computer runs are required.

The *virtual crack extension* method minimises the amount of recomputation required by using a very small change in crack length. The same mesh is used for both problems but co-ordinates of nodes around the crack tip are displaced slightly for the second solution giving a very small δa.

Energy Relationships

The J-integral

The J-integral is an contour integral which is evaluated on a path enclosing the crack tip starting and finishing on the crack faces. Some examples of possible contours are shown in figure 3.14. The two im-

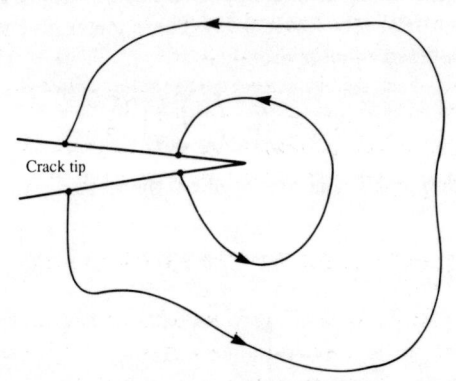

Figure 3.14 Closed contours for J-integrals

portant features of this integral are:
(a) It is path independent i.e. on any of the contours shown in figure 3.14 (or any similar contour) it has the same value.
(b) For linear elasticity the J-integral is equal to the strain energy release rate G and once J is evaluated one of the equations 3.19 and 3.20 can be used to find the SIF.

The J-integral (sometimes called Rice's J-integral) is given by:

$$J = \int_\Gamma \left(U dy - t_i \frac{\partial u_i}{\partial x} ds \right) \tag{3.22}$$

where U is the strain energy density, t_i is the traction vector, u_i is the displacement vector and ds is the element of arc length around the contour Γ. The integral is most often used in the FEM when it is evaluated numerically by a subroutine in the computer program. To avoid any difficulties with the numerical integration the contour should not be too close to the crack tip with its rapidly varying stresses.

The two key advantages of the J-integral method are that only one computer run is required and that the concept is useful in nonlinear analysis (chapter 4). As for the strain energy release rate only single mode problems may be analysed, although other contour integrals have been used to solve mixed-mode problems.

3.7 Green's functions

A Green's function in elasticity usually provides an equation predicting the effects of a point load at a general position. As an example, consider the case of equal and opposite normal point forces P on a crack in an infinite sheet - figure 3.15. These forces give a contribution to the stress intensity factor of:

$$K_I = \frac{P}{\sqrt{\pi a}} \left(\frac{a+b}{a-b} \right)^{1/2} \tag{3.23}$$

If several such point loads are applied then, using the principle of

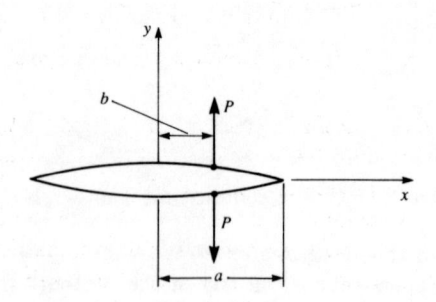

Figure 3.15 Point forces on a crack

superposition, K_I is the sum of the individual contributions. If a continuous load $p(x)$ is applied to the crack surface as in figure 3.16, then the corresponding K_I value is obtained by integrating the Green's function over the crack surface. This gives:

$$K_I = \frac{1}{\sqrt{\pi a}} \int_{-a}^{a} p(x) \left(\frac{a+x}{a-x} \right)^{1/2} dx \tag{3.24}$$

If $p(x)$ is a simple function then the above integral may be performed exactly. If however, $p(x)$ is complicated, or is an approximate function derived from another method, then numerical integration is required. As a first step we can simplify calculations by rescaling the problem so that $a = 1$ and now

$$\begin{aligned} K_I &= \frac{1}{\sqrt{\pi}} \int_{-1}^{1} p(x) \left(\frac{1+x}{1-x} \right)^{1/2} dx \\ &= \frac{1}{\sqrt{\pi}} \int_{-1}^{1} p(x) \frac{(1+x)}{\sqrt{1-x^2}} dx \end{aligned} \tag{3.25}$$

Green's functions

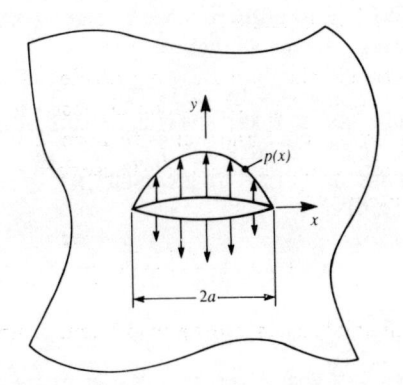

Figure 3.16 A continuously loaded crack

There are some very useful numerical integration methods available called Gauss quadrature (see e.g. Ralston 1965) and we will use one particular form to evaluate equation (3.25).

By choosing the values of x at which we sample the function to be integrated in a special way, very accurate approximations to integrals can be obtained. Gauss quadrature has the remarkable property that if the function being integrated is a polynomial of degree $2n - 1$ then only n points are required to evaluate the integral exactly. The Gauss-Chebyshev variant has the further advantage that the sample points chosen allow for the presence of a $1/\sqrt{1-x^2}$ term in the integrand so that this term is not included in calculations. If we choose x so that:

$$x_j = \cos\frac{(2j-1)\pi}{2n}; \quad j = 1, 2, ..., n \quad (3.26)$$

then

$$\int_{-1}^{1} \frac{1}{\sqrt{1-x^2}} f(x) dx = \frac{\pi}{n} \sum_{j=1}^{n} f(x_j) \quad (3.27)$$

We can now write equation (3.25) as a summation instead of an integral using equation (3.27) so that

$$K_I = \frac{\sqrt{\pi}}{n} \sum_{j=1}^{n} p(x_j)(1+x_j) \quad (3.28)$$

This formula has been used in a BASIC program to compute stress intensity factors from input values of $p(x)$.

Program 3.7 WFI: Evaluates K for a straight crack in an infinite sheet by integrating a Green's function

```
100 PRINT "SIF FOR A CRACK IN INFINITE SHEET. "
110 REM CHEBYSHEV QUADRATURE, STRAIGHT CRACK, CENTRE (0,0)
120 INPUT " HALF CRACK LENGTH A ";A
150 INPUT " NUMBER OF POINTS TO BE USED";N
190 P1=4.0*ATN(1.0)
200 C1=P1/(2*N)
210 REM
220 T=0.0
230 FOR I=1 TO N
240 X=COS((2*I-1)*C1)
250 REM REQUIRES THE NORMAL STRESS AT THE GIVEN X CO-ORDINATE
260 PRINT " AT X = ";X*A;" NORMAL STRESS = ";
270 INPUT S
290 T=T+S*(1+X)
300 NEXT I
310 K1=T/N
330 PRINT " STRESS INTENSITY FACTOR KI [/ SQRT(PI A)] =";K1
340 PRINT
350 INPUT " REPEAT FOR MORE POINTS (Y/N) ";A$
360 IF A$="Y" GOTO 150
370 END
```

Sample run

```
RUN WFI

SIF FOR A CRACK IN INFINITE SHEET
  HALF CRACK LENGTH A ? .2
  NUMBER OF POINTS TO BE USED ? 1
  AT X = -.874228E-08 NORMAL STRESS = ? 1
  STRESS INTENSITY FACTOR KI [ /SQRT(PI A)] = 1

  REPEAT FOR MORE POINTS (Y/N) Y
  Number of points to be used ? 4
  AT X =  .184776 NORMAL STRESS = ? 1
  AT X =  .765367-01 NORMAL STRESS = ? 1
  AT X = -.765367-01 NORMAL STRESS = ? 1
  AT X = -.184776 NORMAL STRESS = ? 1
  STRESS INTENSITY FACTOR KI [ /SQRT(PI A)] = 1
```

```
REPEAT FOR MORE POINTS (Y/N) N
Ready
```

Program notes

(1) The value of x for one integration point should, of course, be zero, the non-zero value occurs because of the limited precision of BASIC arithmetic.

(2) The K-value for a crack in an infinite sheet under a constant normal stress is $\sigma\sqrt{\pi a}$ and it is common practice to normalise SIFs by this quantity. The program result is therefore correct.

(3) Since the function evaluated is linear, its degree is one and the quadrature rule with n points should be exact for polynomials of $2n-1$. Hence with one point the result is exact.

(4) Stress values are only required at discreet points, so it is not necessary to know a function for the stress. Values could be estimated from a graph or table of the stresses on the crack site in the uncracked body.

3.8 Weight functions

A weight function describes how the displacements v on all the boundaries of a cracked body change with the crack length. If a weight function is known for a particular body then stress intensity factors may be obtained for different loadings on the same body by a simple integration. To see how this function relates to the stress intensity factor, we will determine the change in energy introduced by a new crack. Weight functions are discussed in more detail by Parker (1980).

For simplicity, we consider the mode I opening of a crack problem which is symmetric about the x-axis. In figure 3.17(b) the normal stress is assumed to be a known function $-p(x)$ along the dashed line in the uncracked region. When the crack is introduced along the dashed line, the crack surfaces open by a distance v which is a function of x and a. Using the principle of superposition, this crack opening is the same as that produced by a system of equal and opposite forces $p(x)$ applied to the crack surfaces. Now from section 2.2, the change in strain energy U is given by $\frac{1}{2}(force \times extension)$ and hence the change in energy introduced by a varying force and displacement is

(a) Cracked region (b) Uncracked region

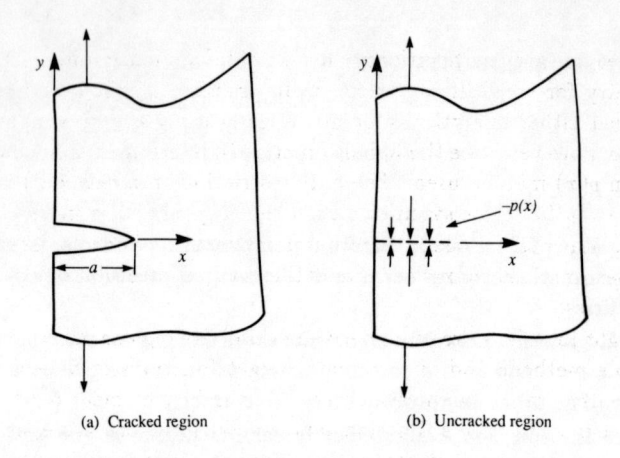

Figure 3.17 Cracked and uncracked regions

the sum of these individual contributions i.e.

$$U = \frac{1}{2}\int_0^a p(x)v(x,a)dx \quad (3.29)$$

We must now relate this energy to the stress intensity factor and from section 3.6 we know that the strain energy density G is given by:

$$G = \frac{K_I^2}{E} = \frac{\partial U}{\partial a} \quad (3.30)$$

for plane stress and $E \Rightarrow E/(1-\nu^2)$ for plane strain. By differentiating equation(3.29) with respect to a

$$\frac{\partial U}{\partial a} = \frac{1}{2}\int_0^a p(x)\frac{\partial v(x,a)}{\partial a}dx \quad (3.31)$$

and substituting from equation(3.30)

$$\frac{K_I^2}{E} = \frac{1}{2}\int_0^a p(x)\frac{\partial v(x,a)}{\partial a}dx \quad (3.32)$$

This last equation is usually arranged in the form:

$$K_I = \int_0^a p(x)m(x,a)dx \quad (3.33)$$

where

$$m(x,a) = \frac{E}{2K_I}\frac{\partial v(x,a)}{\partial a} \quad (3.34)$$

is called the weight function. This weight function is unique to the geometry for which it was derived. In practice, weight functions are obtained either analytically or numerically using a very simple load system; however, once the weight function is determined, any new load system $p(x)$ may be used. The only restriction on a new $p(x)$ is that it must not have less symmetry than the original. For a more general two-dimensional problem, the boundary tractions and displacements will be functions of x and y and the integral is evaluated over all boundaries.

Weight functions are often derived for infinite regions using complex variable methods and approximate weight functions may be derived from numerical techniques such as the boundary element method for finite regions. Green's functions and weight functions are essentially the same in the present context but are derived in a different framework.

3.8.1 Example - *An edge crack in a strip*

We now consider the problem of an infinitely long strip, width W, containing an edge crack of length $a = 1$. Although, in general, weight functions may be determined for loading on all boundaries, in this case we assume that only the crack surface is loaded. The weight function given below is approximate but should be accurate to about 1 percent (Bueckner 1973):

$$m(x) = \frac{\sqrt{1+x}}{\sqrt{2\pi}\sqrt{1-x^2}} \left[1 + m_1(1-x) + m_2(1-x)^2\right] \qquad (3.35)$$

where, if $z = a/W$,

$$m_1 = 0.6147 + 17.1844z^2 + 8.7822z^6$$

$$m_2 = 0.2502 + 3.2889z^2 + 70.0444z^6$$

The integral of the weight function must now be performed over the range 0 to a, as there is only one crack tip. We again use Gauss-Chebyshev quadrature as in the previous section except that integration points are required in the range 0 to 1; this is achieved by changing the constant C1 (line 200) from $\pi/2n$ to $\pi/4n$.

In order to illustrate the use of this new weight function the strip is assumed to be under pure bending, as in figure 3.18. The stress on

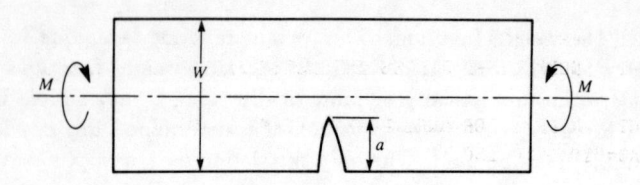

Figure 3.18 Cracked beam under pure bending

the crack site is given by

$$p(x) = (12M/W^3)(W/2 - ax)$$

and this loading is evaluated by line 270 in the program, for simplicity, although pointwise input could be used. The SIF has been normalised by $6M/W^2$ in line 320 so that results may be compared with those given in Rooke and Cartwright directly.

Program 3.8 WFEDG: SIF for edge-cracked strip via weight function.

```
100 PRINT "SIF FOR EDGE CRACK IN INFINITE SHEET WIDTH W."
110 REM CHEBYSHEV QUADRATURE, STRAIGHT CRACK, CENTRE (0,0)
120 INPUT " CRACK LENGTH A ";A
130 INPUT " SHEET WIDTH W ";W
140 INPUT " BENDING MOMENT M ";M
150 INPUT " NUMBER OF POINTS TO BE USED";N
160 Z=A/W
170 M1=0.6147+17.1844*Z*Z+8.7822*Z**6
180 M2=0.2502+3.2889*Z*Z+70.0444*Z**6
190 P1=4.0*ATN(1.0)
200 C1=P1/(4*N)
210 REM INTEGRATION LOOP
220 T=0.0
230 FOR I=1 TO N
240 X=COS((2*I-1)*C1)
250 REM REQUIRES THE NORMAL STRESS AT THE GIVEN X CO-ORDINATE
260 PRINT " AT X = ";X*A;" NORMAL STRESS = ";
270 S=(12*M/W**3)*(W/2-A*X)
280 PRINT S
290 T=T+S*SQRT((1+X)/2)*(1.0+M1*(1.0-X)+M2*(1-X)**2)
300 NEXT I
310 K1=T/N
```

Weight functions

```
320 K1=K1/(6*M/W**2)
330 PRINT ' NORMALISED STRESS INTENSITY FACTOR KI= ';K1
340 PRINT
350 INPUT " REPEAT FOR MORE POINTS (Y/N) ";A$
360 IF A$="Y" GOTO 150
370 END
```

Sample run

```
RUN WFEDG

 SIF FOR EDGE CRACK IN INFINITE SHEET WIDTH W.
  CRACK LENGTH A ? 0.4
  SHEET WIDTH W ? 1
  BENDING MOMENT M ? 1
  NUMBER OF POINTS TO BE USED ? 2
  AT X =  .369552 NORMAL STRESS = ? 1.56538
  AT X =  .153073 NORMAL STRESS = ? 4.16312
  NORMALISED STRESS INTENSITY FACTOR KI = 1.17266

  REPEAT FOR MORE POINTS (Y/N) Y
  NUMBER OF POINTS TO BE USED ? 4
  At X =  .392314 NORMAL STRESS = ? 1.29223
  At X =  .332588 NORMAL STRESS = ? 2.00895
  At X =  .222228 NORMAL STRESS = ? 3.33326
  At X =  .780361E-01 NORMAL STRESS = ? 5.06357
  NORMALISED STRESS INTENSITY FACTOR KI = 1.24063

  REPEAT FOR MORE POINTS (Y/N) N
Ready
```

Program notes

(1) The function being integrated (line 290) includes a non-polynomial term. The integral cannot therefore be evaluated 'exactly', but is evaluated to increasing accuracy as the number of integration points increases. In this case 8 points are sufficient (KI = 1.256) to agree with Rooke and Cartwright to within 1 percent.

It is possible that in some problems the weight function may imply that the crack surfaces overlap and since this is physically impossible, the stress intensity factor will be incorrect. A full discussion of this problem is beyond the scope of this book but note that it can only arise if either $p(x)$ or $m(x,a)$ is negative at some point.

3.9 References

Bueckner, H.F., Field singularities and related integral expressions, in *Methods of analysis and solutions of crack problems*, Ed. G.C. Sih Noordhoff Int., Leiden 1973.

Gallagher, R. H., A review of finite element techniques in fracture mechanics, Proc. 1st Int. Conf. on num. meth. in fracture mechanics, A. R. Luxmore and D. R. J. Owen, Eds. Swansea, 9-13. 1978.

Mason J.C.(1983) *BASIC Numerical Mathematics*, Butterworths, London.

Parker, A.P.(1981) *The Mechanics of Fracture and Fatigue*, E. and F.N. Spon, London.

Ralston, A.(1965) *A First Course in Numerical Analysis*, McGraw-Hill, New York.

Rooke, D.P. and Cartwright, D.J.(1976) *A compendium of stress intensity factors*, HMSO, London.

Smith, R. N. L., From Rags to Riches ? - Developments in the BEM for the solution of elastic fracture problems, Proc. 10th Int. Conf. on boundary elements, C. A. Brebbia (Ed.), Southampton, 155-176, 1988.

Tada, H., Paris, P., and Irwin, G.(1973), *The Stress Analysis of Cracks Handbook*, Del Research Corp., Hellertown, Pennsylvania.

PROBLEMS

(3.1) A flat sheet of length 1 m width 0.5 m has a central crack of length 0.05 m loaded as in figure 3.1 with a stress of $8Pa$ on the ends. What is the stress intensity factor in this case ? (use program 3.2). What would be the percentage error in assuming the infinite plate value ?

(3.2) A plate with height to width ratio h/b=1.0 has a crack of relative width a/b = 0.8 loaded as figure 3.1. Find the configuration correction factor using program 3.2.

(3.3) A plate has with h/b=0.9 has a central crack with a/b = 0.75 Use program 3.2 to find K_I for this geometry if the applied stress is $4Pa$. Hint. first extend data for 2 or 3 curves to a/b = 0.75 .

(3.4) A plate length 2h, width b has an edge crack length a at right angles to a long side and in the centre of it. If $h/b \geq 1.0$ and $a/b \leq 0.6$ then

$$Q = 1.12 - 0.23(a/b) + 10.6(a/b)^2 - 21.7(a/b)^3 + 30.4(a/b)^4$$

Choose 5 values of (a/b) and use program 3.2 to find K_I for several (a/b) values. The 5 values should be chosen for simplicity and do not have to be less than 0.6. Now write a program to evaluate Q directly - which is easier ?

(3.5) A rectangular plate contains a central hole with two radial cracks parallel to the ends of the plate. The stress intensity factors under unit normal tension on the ends of the plate are 1.51 $MPam^{1/2}$. Under a unit pressure on the hole only, the SIFs are 0.2 $MPam^{1/2}$. What is the SIF due to a pin which exerts a uniform pressure on half of the hole when a unit tension is applied to one end ?

(3.6) A plate contains a semicircular cutout on one edge. The stress concentration factor for this case is 6.4. Estimate the SIF for a small crack at the centre of the cutout.

(3.7) A circular hole under biaxial stress has a stress field on the x-axis of the form:

$$\sigma(x) = \sigma_0 \left[1 + \left(\frac{R}{R+x} \right)^2 \right] \quad (3.36)$$

Adapt Program 3.3 to find an approximation for the SIF of an edge crack on the x-axis.

(3.8) A numerical method applied to the solution of a plane strain problem gives the following displacements at the pair of nodes $0.0001m$ from the crack tip on the crack surface.

$u_x = \pm 0.726075E - 6$, $u_y = \pm 0.29043E - 5$.

Find the stress intensity factors if Young's modulus is $200,000 MPa$, Poisson's ratio $\nu = 0.3$ and the crack lies on the x-axis. If the crack were at an angle of 30 degrees to the x-axis how does this change the stress intensity factors ?

(3.9) Values of K_I have been computed for a series of points at a distance r from the crack tip in the table below. What is the best estimate for K_I ?

K	3.20	3.96	3.03	2.17
r	0.01	0.04	0.08	0.12

(**3.10**) Modify the program KDISP to read in the displacements from two pairs of nodes on a quarter-point element and hence compute K. How does the value obtained compare with that from the original KDISP using the data (in metres):

	$r = 0.005$		$r = 0.02$	
u_x	$-0.2714E-6$	$+0.2706E-6$	$-0.5197E-6$	$0.5185E-6$
u_y	$0.8923E-8$	$-0.8899E-8$	$0.1536E-8$	$-0.1583E-8$

with Young's modulus $E = 200,000 MPa$, Poisson's ratio $\nu = 0.3$ and assuming plane stress conditions.

(**3.11**) Use program 3.7 to find the stress intensity factor for a crack in a very large plate if the crack surfaces are loaded with a parabolic stress defined by:
$$p(x) = 1 - x^2$$

(**3.12**) Given that the Green's function for a crack in an infinite sheet under shear stress τ is
$$K_{II} = \frac{1}{\sqrt{\pi a}} \left(\frac{a+x}{a-x} \right)^{1/2}$$
adapt program 3.7 to find both K_I and K_{II}.

(**3.13**) Use program 3.8 to find the SIF for an edge-cracked strip under constant normal tension. What should the result be when the crack is much smaller than the width?

Chapter 4

Crack tip plasticity

4.1 Introduction

Assuming purely elastic behaviour leads to the conclusions of Chapter 2, namely that cracks below the critical length do not grow and that longer cracks cause immediate failure. However, common structural materials do not behave in a perfectly elastic fashion. The high stresses surrounding the crack tip cause plastic deformation and even though this zone may be quite small, it can lead to the step by step growth of very short cracks under cyclic loading. This phenomenon, called fatigue growth, is analysed further in chapter 5, but for the present we concentrate on the plasticity at crack tips which is responsible.

The most important parameter to be determined is the size of the plastic zone, since, if this is small ($\leq a/10$ say), a modified elastic analysis can be used. The additional complexity of the elastic/plastic problem leads us to consider two kinds of approximations of the crack tip region. Firstly, the plastic region is supposed to be of a fixed shape allowing some compensation to be made for the changes produced in the stress field by the plastic deformation. This gives the Irwin and Dugdale models. Secondly, the changes in tip stresses are neglected, giving an approximation to the plastic zone shape. These crack tip plastic zones must be small enough to allow proper plane strain testing of materials. Programs which give the correct testpiece design parameters and estimate material toughness from the tests are given. Some of the problems of even higher levels of plasticity are also mentioned but not considered in great depth.

The analysis in this chapter is restricted to pure crack opening (mode I) behaviour and the changes in material properties caused by plastic flow (work hardening) are neglected. For a more detailed analysis of the topics discussed here see, for example, Ewalds and Wanhill (1984).

4.2 Plastic zone size

The simplest yield criterion predicts plastic failure when the largest principal stress reaches the uniaxial yield stress σ_Y. Since yielding is associated with shear stresses and does not occur under hydrostatic loading, this criterion clearly has some deficiencies. Nevertheless, it does have the advantage of great simplicity ! We can now derive expressions for the size of the plastic zone assuming plane stress conditions.

From chapter 2 we know that if r is small, the normal elastic stress σ is given by:

$$\sigma = \frac{K_I}{\sqrt{2\pi r}} \qquad (4.1)$$

(when $\theta = 0$) and this stress is plotted against the distance r from the crack tip in figure 4.1(a). As a first approximation to the plastic

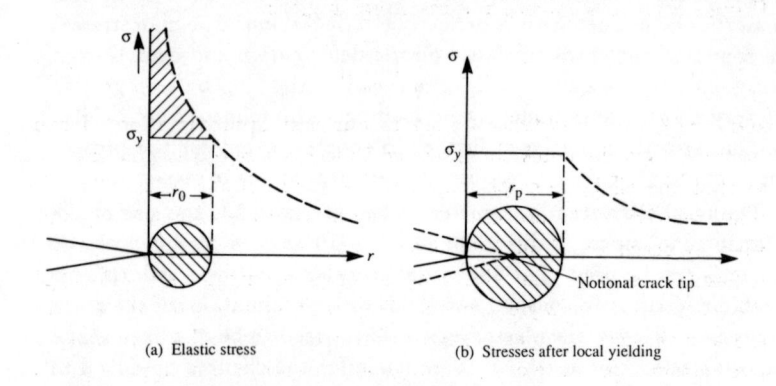

Figure 4.1 Approximations to plastic zones

zone size we could take the width of the region where the elastic stress exceeds the yield stress σ_Y as in figure 4.1(a). Putting $\sigma = \sigma_Y$ and $r = r_0$ in equation 4.1 and rearranging gives:

$$r_0 = \frac{1}{2\pi}\left(\frac{K_I}{\sigma_Y}\right)^2 \qquad (4.2)$$

However, referring again to figure 4.1(a), the shaded area above σ_Y represents a stress which our first approximation has not taken into account. This stress will produce plastic deformation in all the material already stressed to the elastic limit. To obtain an improved approximation, the whole of the elastic force must be balanced by the force required for plastic deformation and this is the basis of the Irwin model.

Now assuming unit thickness, σ is the force per unit length in the x-direction and is given by equation 4.1. The total elastic force is therefore the integral of σ from 0 to r_0 with respect to the distance r. The force per unit length σ_Y required for plastic deformation acts over the length of the plastic zone r_p and hence for equilibrium:

$$\int_0^{r_0} \frac{K_I}{\sqrt{2\pi r}} dr = \sigma_Y r_p \quad (4.3)$$

Therefore:

$$\sqrt{\frac{2}{\pi}} K_I r_0^{1/2} = \sigma_Y r_p \quad (4.4)$$

Using equation 4.2 for r_0 and rearranging gives:

$$r_p = \frac{1}{\pi} \left(\frac{K_I}{\sigma_Y}\right)^2 \quad (4.5)$$

and this estimate is twice as big as our first approximation. Irwin assumed that the crack tip extended to the centre of the plastic zone so that the effective crack length is $a + r_p/2$. If the plastic zone is small relative to the crack, then it may be corrected for by using this effective crack length in SIF calculations.

Dugdale assumed that plastic deformation occurs in a strip ahead of the crack tip - a strip yield model. The crack is supposed to extend all the way through the plastic zone and it is therefore subject to a negative (closing) pressure of σ_Y throughout this zone as in figure 4.2. To derive an expression for zone size, the crack tip singularity is assumed to be exactly cancelled by the negative pressure σ_Y at the extended crack tip. Although the net stress intensity factor is therefore zero, the higher order terms in the crack tip expansions are not and there is thus a finite stress ($= \sigma_Y$) at the crack tip. Assuming that the external stress σ is much less than the yield stress gives an approximation for the length d of the plastic zone

$$d = \frac{\pi}{8} \left(\frac{K_I}{\sigma_Y}\right)^2 \quad (4.6)$$

The Dugdale zone size is a little larger than that predicted by the Irwin model but has the same dependence on K and σ_Y.

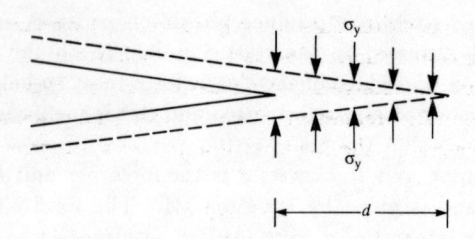

Figure 4.2 The Dugdale model of crack tip zones

4.2.1 Crack tip opening displacement (COD)

The analytical techniques which give the crack tip expansions of chapter 2 may also be used to derive an expression for crack shape under an opening load. For a straight crack in an infinite sheet under constant

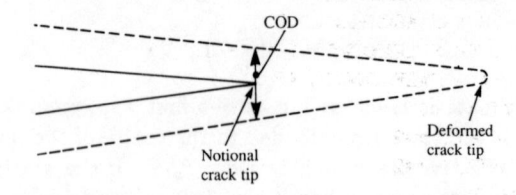

Figure 4.3 Crack opening displacement

normal tension (figure 4.3(a)) the COD δ is

$$\delta = \frac{4\sigma}{E}\left(a^2 - x^2\right)^{1/2} \tag{4.7}$$

assuming plane stress conditions. At the crack tip, $x = a$ and hence $\delta = 0$. However for the Irwin model the effective crack length is $a + r_p/2$. Substituting this effective length into equation 4.7 and neglecting r_p^2 gives

$$\delta = \frac{4\sigma}{E}\left(ar_p\right)^{1/2} \tag{4.8}$$

at $x = a$ and hence from equation 4.5 that

$$\delta = \frac{4}{\pi}\frac{K_I^2}{E\sigma_Y} \tag{4.9}$$

Plastic zone size

At the nominal position of the crack tip ($x = a$) we therefore have a finite opening displacement due to plastic behaviour. The COD may therefore be used to analyse elastic-plastic fracture. Deriving the COD for a Dugdale model requires a more difficult analysis since the crack tip is not opening in the usual fashion; A good approximation (for $\sigma \ll \sigma_Y$) is:

$$\delta = \frac{K_I^2}{E\sigma_Y} \tag{4.10}$$

Estimates of zone size and COD are provided by the following program using both Irwin and Dugdale approximations.

Program 4.2 DUGIN: Plastic zone size and COD

```
100 PRINT "DUGDALE AND IRWIN PLASTIC ZONES (PLANE STRESS)"
110 REM P1=PI, USE CONSISTENT UNITS
120 INPUT " YIELD STRESS ";Y
130 INPUT " YOUNGS MODULUS E";E
140 INPUT " STRESS INTENSITY FACTOR KI ";K
150 INPUT "CRACK HALF LENGTH A";A
160 P1=4.0*ATN(1.0)
170 D=(P1/8)*(K/Y)**2
180 I=(1/P1)*(K/Y)**2
190 DO=K**2/(E*Y)
200 IO=(4/P1)*K**2/(E*Y)
210 PRINT
220 PRINT "                    DUGDALE                IRWIN"
230 PRINT " ZONE SIZE       ";D;"             ";I
240 PRINT "EFFECTIVE LENGTH ";(A+D);"           ";(A+I/2)
250 PRINT "     COD         ";DO;"           ";IO
260 END
```

Sample run

```
RUN DUGIN

DUGDALE AND IRWIN PLASTIC ZONES (PLANE STRESS)
  YIELD STRESS ? 500
  YOUNGS MODULUS E ? 200000
  STRESS INTENSITY FACTOR KI   ? 40
CRACK HALF LENGTH A ? .04
```

```
                  DUGDALE              IRWIN
   ZONE SIZE      .251327E-02          .203718E-02
EFFECTIVE LENGTH  .425133E-01          .410186E-01
        COD       .000016              .203718E-04
```

Ready

Program notes

(1) The Irwin zone size is less than the Dugdale zone size, but the reverse is true for the COD.
(2) The plastic zone size does not depend on E and if this is the only quantity required any number can be input for E ((1 say) although the COD will now be incorrect.

4.3 The shape of the plastic zone

In the previous section a very simple yield criterion is used to redistribute stresses and derive estimates of plastic zone size. We now consider two more sophisticated yield criteria and use them to obtain approximations to plastic zone shape - but without redistribution of stresses to obtain equilibrium. Yield criteria are usually written in terms of principal stresses $\sigma_1 > \sigma_2 > \sigma_3$.

The Tresca criterion
The maximum shear stress at a point is $(\sigma_1 - \sigma_3)/2$. The Tresca criterion assumes that yielding will occur whenever the maximum shear stress is greater than the maximum shear stress in uniaxial tension ($\sigma_Y/2$ where σ_Y is the yield strength of the material). Yielding just occurs when

$$\sigma_1 - \sigma_3 = \sigma_Y \qquad (4.11)$$

The Von Mises criterion
If the maximum shear strain energy per unit volume is equal to the corresponding energy which causes yielding in uniaxial tension then yielding will occur according to the Von Mises criterion. Mathematically, yielding occurs when

$$(\sigma_1 - \sigma_2)^2 + (\sigma_2 - \sigma_3)^2 + (\sigma_3 - \sigma_1)^2 = 2\sigma_Y^2 \qquad (4.12)$$

The shape of the plastic zone

Principal stresses

The Von Mises and the Tresca criteria both predict yielding in terms of principal stresses. Taking the first term of the crack tip expansions in chapter 2 and reformulating as principal stresses as in section 2.1 gives

$$\sigma_1 = \frac{K}{(2\pi r)^{1/2}} \cos\left(\frac{\theta}{2}\right) \left[1 + \sin\left(\frac{\theta}{2}\right)\right]$$

$$\sigma_2 = \frac{K}{(2\pi r)^{1/2}} \cos\left(\frac{\theta}{2}\right) \left[1 - \sin\left(\frac{\theta}{2}\right)\right] \quad (4.13)$$

while for plane strain

$$\sigma_3 = \nu(\sigma_1 + \sigma_2) = 2\nu \frac{K}{(2\pi r)^{1/2}} \cos\left(\frac{\theta}{2}\right)$$

and for plane stress

$$\sigma_3 = 0$$

Zone shape

Substituting the above equations for principal stresses into Von Mises' criterion and rearranging gives the following expressions for the extent r of the plastic region. For plane stress:

$$r = \frac{K^2}{2\pi \sigma_Y^2} \cos^2\left(\frac{\theta}{2}\right) \left[1 + 3\sin^2\left(\frac{\theta}{2}\right)\right] \quad (4.14)$$

and for plane strain:

$$r = \frac{K^2}{2\pi \sigma_Y^2} \cos^2\left(\frac{\theta}{2}\right) \left[(1 - 2\nu)^2 + 3\sin^2\left(\frac{\theta}{2}\right)\right] \quad (4.15)$$

Note that it makes no difference in the Von Mises' criterion if the principal stresses are not ordered correctly in size since all combinations appear. However, for the Tresca criterion we require the largest and the smallest principal stresses and equations (4.13) change in relative magnitude as θ changes. For the case of plane stress this causes no difficulty since $\sigma_3 = 0$. In the plane strain case there are two equations for r based on $\sigma_1 - \sigma_2$ and $\sigma_1 - \sigma_3$ and the largest modulus must be selected. For plane stress:

$$r = \frac{K^2}{2\pi \sigma_Y^2} \cos^2\left(\frac{\theta}{2}\right) \left[1 + \sin\left(\frac{\theta}{2}\right)\right]^2 \quad (4.16)$$

and for plane strain:

$$r = \frac{K^2}{2\pi\sigma_Y^2} \cos^2\left(\frac{\theta}{2}\right) \left[1 - 2\nu + \sin\left(\frac{\theta}{2}\right)\right]^2 \quad (4.17)$$

or

$$r = \frac{K^2}{2\pi\sigma_Y^2} \cos^2\left(\frac{\theta}{2}\right) \left[2\sin\left(\frac{\theta}{2}\right)\right]^2 \quad (4.18)$$

whichever is the larger.

By calculating the extent r of the plastic zone for a range of angles θ the shape of this zone can be plotted as in figure 4.4. The key fea-

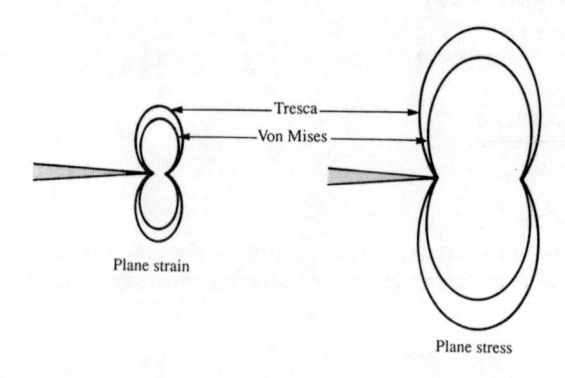

Figure 4.4 Plastic zone shapes for plane strain and plane stress

ture of the zone shapes is that the plane strain plastic zone is much smaller than that for plane stress. The two yield criteria give broadly similar results with the Tresca zone always larger than or equal to the Von Mises zone. The various equations for zone size are incorporated in a BASIC program ZONE which gives estimates of zone size r corresponding to input parameters.

Program 4.3 ZONE: plastic zone size

```
100 PRINT " PLASTIC ZONE SIZE, VON MISES AND TRESCA CRITERIA"
110 REM P1=PI, N=POISSONS RATIO (NU)
120 P1=4.0*ATN(1.0)
130 N=0.3
140 INPUT " KI ";K
150 INPUT " YIELD STRESS ";Y
```

```
160 C=K**2/(2*P1*Y**2)
170 INPUT " PLANE STRESS (Y/N)",A$
180 INPUT " ANGLE THETA (DEGREES) ";T
190 IF T>360 GOTO 380
200 T=T*P1/180
210 C2=COS(T/2)
220 S2=SIN(T/2)
230 C=C*C2**2
240 IF(A$="Y") GOTO 310
250 T$="PLANE STRAIN"
260 R1=C*(3*S2**2+(1-2*N)**2)
270 R2=C*(1-2*N+S2)**2
280 R3=C*(4*S2**2)
290 IF R3>R2 THEN R2=R3
300 GOTO 340
310 T$="PLANE STRESS"
320 R1=C*(1+3*S2**2)
330 R2=C*(1+S2)**2
340 PRINT T$
350 PRINT " VON MISES";R1;"   TRESCA";R2
360 PRINT
370 GOTO 180
380 END
```

Sample run 1

```
RUN ZONE

 PLASTIC ZONE SIZE, VON MISES AND TRESCA CRITERIA
 KI ? 100
 YIELD STRESS ? 500
 PLANE STRESS (Y/N)       ? Y
 ANGLE THETA (DEGREES) ? 0
PLANE STRESS
 VON MISES .63662E-02   TRESCA .63662E-02

 ANGLE THETA (DEGREES) ? 30
PLANE STRESS
 VON MISES .713341E-02   TRESCA .941227E-02

 ANGLE THETA (DEGREES) ? 900
```

Sample run 2

RUN ZONE

```
 PLASTIC ZONE SIZE, VON MISES AND TRESCA CRITERIA
 KI ? 100
 YIELD STRESS ? 500
 PLANE STRESS (Y/N)      ? N
 ANGLE THETA (DEGREES) ? 0
PLANE STRAIN
 VON MISES .101859E-02   TRESCA .101859E-02

 ANGLE THETA (DEGREES) ? 30
PLANE STRESS
 VON MISES .214402E-02   TRESCA .25781E-02

 ANGLE THETA (DEGREES) ? 900
Ready
```

Program notes

(1) Since all the formulae include the term $K^2 \cos^2(\theta/2)/2\pi\sigma_Y^2$ it is precomputed in lines 160 and 230.

(2) To allow the input of positive and negative angles between 0 and 360 degrees, the program terminates if θ is greater than 360 degrees.

4.4 Elastic-plastic failure

The previous section attempted to show some approximations to the size and shape of plastic crack tip zones. The process of crack growth with significant plastic deformation is still too difficult to express mathematically with any great precision, despite the use of much more complicated models than those considered here. We will therefore attempt a qualitative description of some fracture phenomena.

Considering fracture in plates of finite thickness there are three ways in which plastic behaviour may be modelled: plane stress, plane strain and a transitional state between the two extremes. Plane stress is normally assumed if the calculated size of the plane stress fracture zone (r_p for Irwin) is of the same order as the plate thickness. Plane strain is assumed if the calculated plane stress r_p is about 10 per cent or less of the plate thickness. It is worth noting that the plane

Elastic-plastic failure

strain zone size tends to be larger than that predicted in section 4.2, whilst the plane stress zone tends to be smaller than predicted. For a thick plate containing a plane crack a model of plastic zone shape is of the form shown in figure 4.5. At the plate surface there can

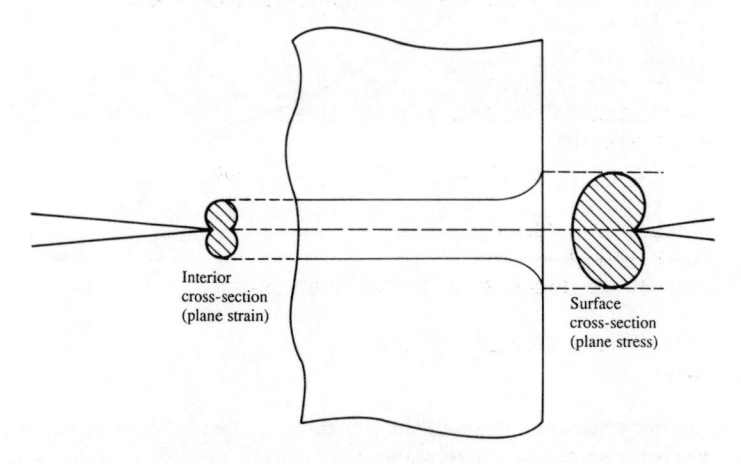

Figure 4.5 Model of the plastic zone in a thick plate

be no normal stress and hence a state of plane stress exists with a correspondingly large plastic zone. Deep in the interior of the plate there is a plane strain state which becomes transitional in a layer near the crack surface.

4.4.1 Types of fracture

In this section we consider ways in which common structural metals like steel and aluminium may actually fail at moderate temperatures. Suppose that a cracked plate of such a material is loaded until it breaks. The nature of the breakage and the amount of stress required is a function of the plate thickness.

A thin plate will be under predominantly plane stress conditions ($\sigma_3 = 0$) and from say, the Von Mises criterion, a high level of shear stress is possible. For pure plane stress, the maximum shear stress occurs on planes at 45 degrees to the crack surface. A thin plate therefore usually fails with significant plastic deformation on an angled surface as shown in figure 4.6. This is known as *ductile shear failure*

A thick plate has a smaller plastic zone around the crack tip and

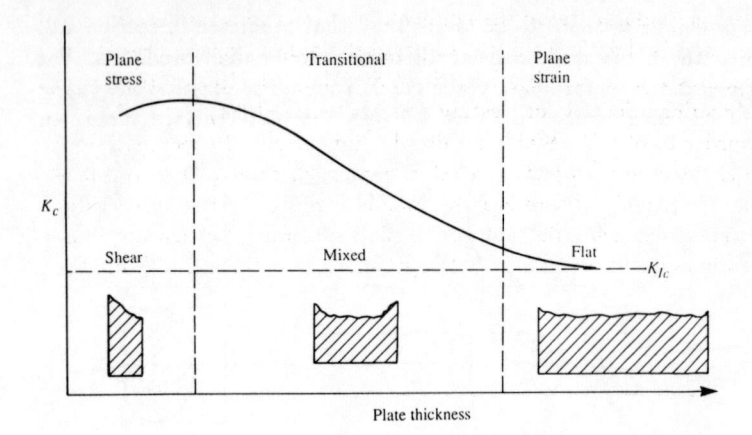

Figure 4.6 Types of failure

supports a stress σ_3 through the thickness of the plate. Failure due to shear stresses is now unlikely since the terms in the Von Mises criterion say, are now all relatively small. The plate thus fails by cleavage of the material in the crack plane producing a flat fracture. This type of failure is known as *brittle failure*.

For plates of intermediate thickness, aspects of both brittle and ductile failure occur. In the central region which is under plane strain conditions the fracture is flat but near the edge of the plate shear stresses dominate resulting in a slanting edge or *shear lip* as in figure 4.6.

Figure 4.6 also illustrates the variation of stress intensity factor with plate thickness and hence the degree of plasticity. The value of K at failure is called the *fracture toughness* and the lower limiting value K_{Ic}, the *plane strain fracture toughness*. In practice, the resistance of material types to failure due to cracking is measured experimentally and quoted in terms of K_{Ic}. This is because K_{Ic} provides a conservative bound on the fracture toughness and is a material constant, whereas failure in a mixed stress regime is a function of material thickness.

4.5 Plane strain fracture toughness

In the last section the lower limiting value of stress intensity factor is stated to be the value obtained from plane strain testing. This testing

Plane strain fracture toughness

process is fundamental to the application of fracture considerations to design and is therefore subject to precisely laid out procedures. Such procedures are defined by the British Standards Institute (1977) and the American Society for Testing and Materials (1984).

In order to obtain reliable results the dimensions of specimens used in such tests are subject to a set of empirical rules to ensure plane strain conditions. There are two standard designs - the three point bend specimen and the compact tension specimen - which are illustrated in figures 4.7 and 4.8. Both specimens have thickness B and the

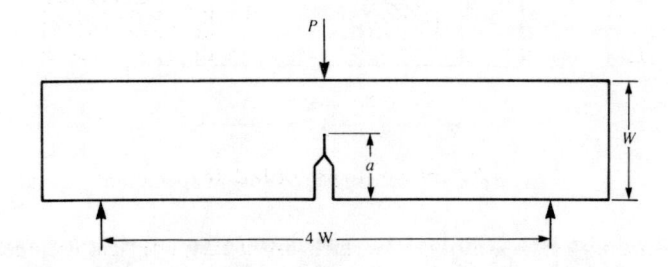

Figure 4.7 The three point bend specimen

effective crack length a should be between 0.45 and 0.55 of W. The dimensions which should give plane strain conditions are as follows:

$$a \geq 2.5 \left(\frac{K_{Ic}}{\sigma_Y}\right)^2$$

$$B \geq 2.5 \left(\frac{K_{Ic}}{\sigma_Y}\right)^2 \quad (4.19)$$

$$W \geq 5.0 \left(\frac{K_{Ic}}{\sigma_Y}\right)^2$$

Thus, before designing the specimen, it is necessary to estimate K_{Ic}! This is usually done from experience of testing similar materials and using an overestimate. Once the specimen has been manufactured it is necessary to produce a very sharp crack at the machined notch in the specimen - precracking. This is done by applying an oscillating load which causes a fatigue crack (see Chapter 5). The maximum fatigue load should not exceed 60 percent of the final breaking load.

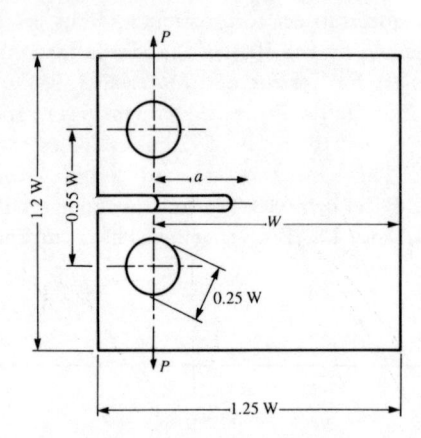

Figure 4.8 The compact tension specimen

The next step is to load the specimen to failure with a slowly increasing load (K increasing at 0.55 to 2.75 $MPa\ m^{1/2}/s$). The load and displacement are monitored to produce load-displacement curves of the form shown in figure 4.9. Since load-displacement curves vary in shape near failure a standard procedure is followed to determine the final load. A tangent to the curve is drawn from the origin and a second line whose gradient is 5 percent less is also added. The load at failure P_Q is taken to be the highest point between these two lines and is indicated in figure 4.9. If the maximum recorded load is greater than $1.1P_Q$, too much yielding has occured and the testpiece must be redesigned.

Values of stress intensity factors may be found by using experimentally validated formulae for test specimens - equations (4.20) and (4.21). For convenience we define $q = a/W$ and then for the bend specimen:

$$K = \frac{6P\sqrt{a}}{BW}\left[1.93 - 3.07q + 14.53q^2 - 25.11q^3 + 25.8q^4\right] \qquad (4.20)$$

While for the tension specimen

$$K = \frac{P\sqrt{a}}{BW}\left[29.6 - 185.5q + 655.7q^2 - 1017q^3 + 638.9q^4\right] \qquad (4.21)$$

In particular, to determine K_{Ic}, the load P is taken as the failure load P_Q.

Plane strain fracture toughness

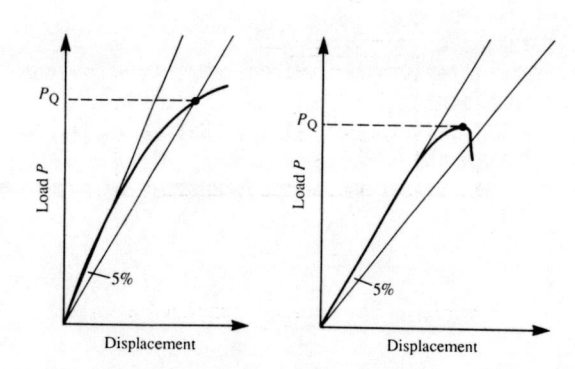

Figure 4.9 Load-displacement curves to failure

A program which determines the fracture toughness from input test data is given below. The program also checks the failure load and the testpiece dimensions for consistency with the testing guidelines.

Program 4.5 CTS: Fracture toughness using a compact tension specimen

```
100 PRINT " FRACTURE TOUGHNESS OF COMPACT TENSION SPECIMEN"
110 INPUT " THICKNESS B(MM)";B
120 B=B/1000
130 INPUT " WIDTH TO HOLE CENTRES W(MM)";W
140 W=W/1000
150 INPUT " CRACK LENGTH TO HOLE CENTRES A(MM)";A
160 A=A/1000
170 INPUT " LOAD AT FAILURE P(KN)";P
180 INPUT " MAXIMUM RECORDED LOAD L(KN)";L
190 IF L<1.1*P GOTO 220
200 PRINT " LOAD EXCEEDS 1.1 PQ - TEST INVALID"
210 GOTO 360
220 P=P/1000
```

```
230 INPUT " MATERIAL YIELD STRENGTH Y (MPA)";Y
240 Q=A/W
250 K=P*SQRT(A)/(B*W)
260 K=K*(29.6-185.5*Q+655.7*Q**2-1017*Q**3+638.9*Q**4)
270 PRINT
280 PRINT " FRACTURE TOUGHNESS KIC";K
290 T=2.5*(K/Y)**2
300 IF A>T THEN 320
310 PRINT " TEST INVALID A=";A*1000;"MUST BE >";T*1000
320 IF B>T THEN 340
330 PRINT " TEST INVALID B=";B*1000;"MUST BE >";T*1000
340 IF W>2*T THEN 360
350 PRINT " TEST INVALID W=";W*1000;"MUST BE >";2*T*1000
360 END
```

Sample run 1

```
RUN CTS

 FRACTURE TOUGHNESS OF COMPACT TENSION SPECIMEN
 THICKNESS B(MM) ? 25
 WIDTH TO HOLE CENTRES W(MM) ? 80
 CRACK LENGTH TO HOLE CENTRES A(MM) ? 40
 LOAD AT FAILURE P(KN) ? 36
 MAXIMUM RECORDED LOAD L(KN) ? 38.5
 MATERIAL YIELD STRENGTH Y (MPA) ? 500

 FRACTURE TOUGHNESS KIC ? 48.8925
Ready
```

Plane strain fracture toughness

Sample run 2

```
RUN CTS

FRACTURE TOUGHNESS OF COMPACT TENSION SPECIMEN
THICKNESS B(MM) ? 25
WIDTH TO HOLE CENTRES W(MM) ? 80
CRACK LENGTH TO HOLE CENTRES A(MM) ? 40
LOAD AT FAILURE P(KN) ? 36
MAXIMUM RECORDED LOAD L(KN) ? 38.5
MATERIAL YIELD STRENGTH Y (MPA) ? 370

FRACTURE TOUGHNESS KIC ? 48.8925
TEST INVALID A= 40 MUST BE > 43.6537
TEST INVALID B= 25 MUST BE > 43.6537
TEST INVALID W= 80 MUST BE > 87.3074
Ready
```

Program notes

(1) For convenience data is input in mm and KN and converted within the program. It is assumed that dimensions fall within the accepted ranges shown in figure 4.8.

(2) The fracture toughness is calculated in lines 250 and 260. To adapt the program for the three-point bend specimen the appropriate formulae may be inserted here.

(3) Sample run 1 and sample run 2 are identical except that in run 2 the material has a lower yield strength which increases the size of the plastic zone sufficiently to invalidate the test.

4.5.1 Testpiece design

Program 4.5 checks that the final fracture load is within accepted tolerances for plane strain conditions to apply. In order to design a suitable specimen we first need a value for the yield stress and an estimate of the maximum expected fracture toughness. This will enable us to choose a suitable specimen thickness using equation (4.19).

It is also important to estimate the maximum load which is likely to occur since this must be within the capacity of the available testing machinery. Although the bend specimen requires about 2.5 times as much material as the compact tension specimen, it should break at a lower load level (say 10 percent less). When the maximum load anticipated for a compact tension specimen exceeds testing capacity, a bend specimen must be used. If a longer than standard bend specimen is used to reduce the maximum load further, then equation (4.20) must be modified to take account of this.

The following program for the design of compact tension testpieces assumes that $W = 2B$ and prints a list of dimensions corresponding to the input yield stress and (estimated) fracture toughness. By rearranging equation (4.21) to obtain an expression for the load P using the given value of K, the failure load can be computed. This failure load is, of course, a function of the crack length a which is not known before fatigue precracking. A 'worst case' analysis assumes the shortest allowable crack ($a = 0.45W$) which requires the largest load to produce failure. If load levels are critical then precracking could be controlled to ensure that crack length is near the upper limit, thus reducing maximum load levels.

The program also allows the input of convenient nominal dimensions greater than those initially computed if desired.

Program 4.5.1 DCTS: Design of the compact tension specimen

```
100 PRINT " DESIGN OF COMPACT TENSION SPECIMEN"
110 PRINT " (DIMENSIONS GIVEN IN MM)"
120 INPUT " MATERIAL YIELD STRENGTH Y (MPA)";Y
130 INPUT " MAXIMUM FRACTURE TOUGHNESS KIC MNM-3/2";K
140 B=2.5*(K/Y)**2
150 W=2*B
160 A=0.45*W
170 PRINT
180 PRINT " THICKNESS B =";B*1000
190 PRINT " HEIGHT 1.2W =";1.2*W*1000
200 PRINT " OVERALL WIDTH 1.25W =";1.25*W*1000
210 PRINT " HOLE CENTRES AT 0.55W =";0.55*W*1000
220 PRINT " HOLE DIA 0.25W =";0.25*W*1000
230 PRINT
240 Q=A/W
250 P=K*B*W/SQRT(A)
260 P=P/(29.6-185.5*Q+655.7*Q**2-1017*Q**3+638.9*Q**4)
270 PRINT " IF A=";Q;"W OR";A*1000;"MM"
280 PRINT " ESTIMATED LOAD REQUIRED (KN)";P*1000
290 PRINT
300 INPUT "THICKNESS B(MM) (ZERO TO FINISH)";B
310 IF B=0 GOTO 370
```

```
320 B=B/1000
330 W=2*B
340 INPUT "CRACK LENGTH A AS A MULTIPLE OF W";A
350 A=A*W
360 GOTO 170
370 END
```

Sample run 1

```
RUN DCTS

  DESIGN OF COMPACT TENSION SPECIMEN
  (DIMENSIONS GIVEN IN MM)
  MATERIAL YIELD STRENGTH Y (MPA) 700
  MAXIMUM FRACTURE TOUGHNESS KIC MPA M^1/2 110

  THICKNESS B = 53.7778
  HEIGHT 1.2W = 129.067
  OVERALL WIDTH 1.25W = 134.444
  HOLE CENTRES AT 0.55W = 59.1555
  HOLE DIA 0.25W = 26.8889

  IF A= 0.45 W OR 48.4 MM
  ESTIMATED LOAD REQUIRED (KN) 232.685

THICKNESS B(MM) (ZERO TO FINISH) 55
CRACK LENGTH A AS A MULTIPLE OF W .45

  THICKNESS B = 55
  HEIGHT 1.2W = 122
  OVERALL WIDTH 1.25W = 137.5
  HOLE CENTRES AT 0.55W = 60.5
  HOLE DIA 0.25W = 27.5

  IF A= 0.45 W OR 49.5 MM
  ESTIMATED LOAD REQUIRED (KN) 240.663

THICKNESS B(MM) (ZERO TO FINISH) 0
Ready
```

Program notes

(1) All calculations use dimensions in metres, although they are input and output in terms of mm.
(2) If the first estimate of peak load is well within testing machine capacity, dimensions can be rounded up to some convenient figure. The peak load is automatically calculated for the new dimensions.
(3) If the peak load slightly exceeds machine capacity then a crack length nearer to $0.55W$ may be assumed, though this may require very careful control of the precracking. If the peak load still exceeds machine capacity, a bend specimen must be considered.

4.6 Failure at higher levels of plasticity

Thus far, it has been assumed that failure due to fracture can be adequately modelled using a linear elastic approach with corrections for small plastic zones around the crack tips. There are many applications in which more extensive plastic deformation may occur. As an example, the determination of fracture toughness using standard specimens requires that these specimens are large enough to ensure sufficiently small plastic zones. This may make such specimens rather unwieldy or of a size that prohibits testing on available machines. However, the use of smaller specimens may give such large plastic zones that assumptions of linear elastic behaviour are completely invalidated.

There is therefore a need for a model which can cope with more extensive plasticity than the modified stress intensity approach, both in design and testing applications. Note that in contrast to elastic failure, there may be a period of stable crack growth before breakage. Two models for fracture in the presence of moderate plasticity are outlined below. It is still assumed that plasticity is not sufficiently extensive to give general yielding and thus eventual plastic collapse.

4.6.1 Crack tip opening displacement (COD)

In section 4.1 we saw that if a crack deforms plastically at the tip, then instead of zero displacement, a displacement δ occurs. For the modified linear elastic model, δ can be expressed in terms of the stress intensity factor. However, for more extensive plastic deformation the stress intensity factor approach is no longer valid, so an empirical method based on the crack opening displacement has been developed.

Failure at higher levels of plasticity

The COD test uses a bend specimen which is subject to a British Standard in the same way as for fracture toughness testing (BS 1979). A load-displacement curve is plotted and the value of COD estimated. Failure is unlikely to be an instantaneous event as in more elastic testing so that δ_{crit} may be defined as the value at one of three points; The beginning of stable crack extension, the point of instability or the maximum recorded load.

Once a critical value of COD has been obtained, then the maximum permissible crack size in a particular cracked body is determined from the empirical formulae:

$$a_{max} = \frac{\delta_{crit} E \sigma_Y}{2\pi \sigma_1^2} \qquad (4.22)$$

if $\sigma_1/\sigma_Y \leq 0.5$, while for $0.5 \leq \sigma_1/\sigma_Y \leq 2$

$$a_{max} = \frac{\delta_{crit} E}{2\pi \sigma_1 - 0.25\sigma_Y} \qquad (4.23)$$

where σ_1 is the total of all stress components including any residual stresses.

The above COD model is a conservative one, usually predicting a maximum allowable crack size below that producing the critical COD. The COD approach has been extensively applied to the analysis of cracks in welded structures.

4.6.2 The J-integral

The use of the J-integral for linear elastic materials is described in chapter 3, and is equivalent to the strain energy release rate G only for linear elasticity. J is defined as the rate of change of potential energy U per unit thickness with respect to crack length a, that is

$$J = -\frac{\partial U}{\partial a} \qquad (4.24)$$

Now relationships involving J and potential energy are still valid if we consider purely elastic behaviour, even if the relationship is non-linear. In particular, we can take non-linearnon-linear elasticity elastic behaviour under monotonic loading to be identical to that for the elastic-plastic material being analysed. A typical load-extension curve for an uncracked plate is shown in figure 4.10 and this figure can represent either non-linear elasticity or elastic-plastic deformation under

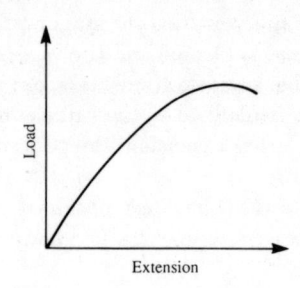

Figure 4.10 Load-extension curve for non-linear elasticity

increasing load. If the load is reduced the plate will behave differently according to the nature of the material. A non-linear elastic material will return to its original shape and follow the same curve under loading and unloading. An elastic-plastic material will have some region of plastically deformed material which will cause the load-extension curve to follow a different path on unloading. If the J-integral is used for non-linear elastic modelling of elastic-plastic material it is therefore invalid for any unloading case. Since the crack tip is unloaded when crack growth occurs it is theoretically invalid for this situation.

As in chapter 3, J may be expressed in terms of a contour integral around an unloaded crack tip and its value should be independent of the contour used. Now we must impose an additional condition - that the contour is outside the region of plastic deformation. The integral may be written as follows:

$$J = \int_\Gamma \left(U dy - t_i \frac{\partial u_i}{\partial x} ds \right) \quad (4.25)$$

where U is the strain energy density, t_i is the traction vector, u_i is the displacement vector and ds is the element of arc length around the contour Γ. Using a numerical approach such as the finite element method, solutions for the stresses and displacements in a cracked body can be determined and hence a value for J. The above path-independent integral expression for J assumes plane strain conditions.

As for previous fracture parameters, we assume that there is a critical value of J, J_c, which determines the beginning of crack growth in a given material. This critical value must be determined from experimental measurements in a broadly similar fashion to K_{Ic}. Procedures to be followed for three-point bend and compact tension specimens have been defined by the American Society for Testing and Materials (1982). Once J_c has been determined experimentally for a given material, the J value for a cracked component is usually obtained numerically and crack growth assumed to occur if $J \geq J_c$.

4.6 References

Ewalds, H.L., and Wanhill, R.J.H. (1984) fracture mechanics, Edward Arnold, London.

ASTM Standard E 399-83, (1984) *Standard Test Method for Plane-strain Fracture Toughness Testing of Metallic Materials*, 1984 book of ASTM standards, Philadelphia.

B.S. 5447 (1977) *Methods for Plane Strain Fracture Toughness Testing*, British Standards Institute, London.

B.S. 5762 (1979) *Methods for Crack Opening Displacement (COD) Testing*, British Standards Institute, London.

ASTM Standard E 813-81 (1982) *Standard Test Method for J_{Ic}, A Measure of Fracture Toughness.*, 1982 book of ASTM standards, Philadelphia.

PROBLEMS

(4.1) A thin steel sheet has a yield stress of $400MPa$ and contains a crack whose SIF is $75MPa\ m^{1/2}$. How big is the plastic zone around the crack tip ? What happens to the size of this zone if the the yield stress is halved ? What is the effect on zone size if the crack length is halved ?

(4.2) A thin sheet of material with Young's modulus $210,000MPa$ and yield stress $450MPa$ contains a crack with SIF=$45MPa\ m^{1/2}$. Assuming a Dugdale model how much should the crack open at the tip ?

(4.3) Derive the equations for the Von Mises and Tresca plastic zone sizes given in section 4.3.

(4.4) Use program 4.3 ZONE to obtain the extent of the plastic zone around a crack tip for a range of angles between 0 and 360 degrees. Plot the shapes of the zones under each criteria and compare them with figure 4.4. What happens to the size of these zones if K is halved ?

(4.5) Use program 4.3 to study how the shapes of the crack tip plastic zones under the Von Mises and Tresca criteria are affected by changes in Poisson's ratio.

(4.6) A plate of material with yield stress $500 MPa$ has a crack with a theoretical stress intensity factor of $75 MPa\ m^{1/2}$. What type of stress condition should be assumed if the plate is (a) 10 cm thick (b) 1 cm thick ? Assuming an Irwin type model what would you take as the effective stress intensity factor ?

(4.7) A compact tension specimen has the following dimensions: $B = 25.1 mm$, $W = 50.1 mm$, $a = 24.3 mm$. If the failure load is $39.5 MN$, the maximum recorded load is $42 MN$ and the yield stress is $720 MPa$, is the test valid ? What is K_{Ic} ?

(4.8) A compact tension specimen has dimensions as follows: $B = 49.8 mm$, $W = 100.1 mm$, $a = 50.3 mm$. If the failure load is $234 MN$, the load is a maximum of $255 MN$ and the yield stress is $970 MPa$, is the test valid ? What is K_{Ic} ?

(4.9) Modify program CTS to analyse the results of a three-point bend test. A bend specimen has dimensions $B = 25 mm$, $W = 50 mm$, $a = 24.2 mm$; If the maximum load and load at failure is $25.6\ kN$ and proof stress is $1280\ MPa$ can you determine a valid K_{Ic} ?

(4.10) Write a program which inputs yield stress and an estimated value of fracture toughness and outputs the required testpiece dimensions for a three-point bend specimen. Compare your results with the corresponding values for a compact tension specimen.

Chapter 5

Crack growth

5.1 Introduction

In chapter 2 we saw that a crack in a perfectly elastic material will normally return to its original shape when an applied loading is removed. The exception to this rule occurs if the loading exceeds a critical value, in which case the crack grows suddenly and catastrophically. However, practical engineering materials do not behave perfectly elastically but undergo plastic deformation in some, perhaps very small, region around the crack tip and this phenomenon has been described in chapter 4. Thus, in practical materials, the crack does not return to its original shape when the applied load is removed but some permanent change has taken place. A sequence of increasing and decreasing loads (cyclic loading) can therefore lead to an increase in crack length at each step - crack growth by fatigue. This type of crack growth takes place at loads far below that required for sudden failure but the rate of growth increases as the crack becomes larger, until the crack is large enough for total failure to occur. Using experimentally derived laws and linear elastic fracture mechanics we can predict the rate of crack growth for materials with moderate plasticity, and hence a safe life for components at the design stage.

If cracks are simply loaded at right angles to the crack plane they will usually grow in the same plane, although some crack paths may be unstable due to second order effects at the crack tip. Second order terms are described in detail since they also affect the direction and onset of stable crack growth although the exact relationship is not always clear. If loading is of mixed-mode type then predicting the direction in which cracks will grow is just as difficult as deciding how fast they will do so. In sections 5.4, 5.5 and 5.6 we discuss two commonly used criteria which give a fair approximation to experimentally observed behaviour in both fast fracture and in fatigue.

5.2 Fatigue crack growth

Fatigue growth occurs because of the inelastic behaviour at the crack tip. Taking the Irwin model from chapter 4, the radius of the circular plastic zone is

$$r_z = \frac{1}{2}r_p = \frac{1}{2\pi}\left(\frac{K}{\sigma_p}\right)^2 \tag{5.1}$$

When the load is increased, K increases, deforming the zone around the crack tip plastically. When the load decreases, the crack tip stresses do not reduce to zero because of the deformed zone around the crack tip. A reduction in load may therefore result in further deformation. This process is called *reversed yielding* and takes place in a zone of about one-quarter of r_z in radius. Under cyclic loading the crack grows into the reversed yielding zone, the plastic zone then extends further and the whole process is repeated.

Experimentally, we can take a cracked specimen and apply a stress σ_a. The stress is then increased to σ_b, then reduced to σ_a, then increased to σ_b, and so on. This type of loading is called *constant amplitude cyclic loading* since the size of the change in stress $\Delta\sigma$ is a constant. Repeating the experiment on a similar cracked specimen but with a different amplitude of loading leads to a different crack growth rate. A typical series of graphs is shown in figure 5.1. As

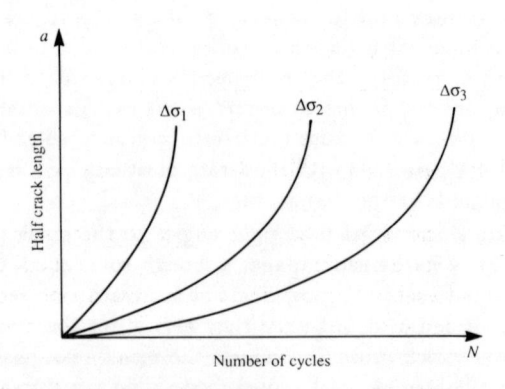

Figure 5.1 Crack growth rate for different load amplitudes

one might expect, the highest amplitude loading produces the highest growth rate, but to obtain a simple relationship some manipulation is

Fatigue crack growth

needed. Firstly, the load amplitude is usually expressed in terms of the effect on the crack, that is in terms of stress intensity factors. The stress intensity factor range ΔK is

$$\Delta K = K_{max} - K_{min} = Q\Delta\sigma(\pi a)^{1/2} \qquad (5.2)$$

The quantity of interest is the rate at which the crack grows as the number of cycles is increased. This rate, da/dN is the gradient of the curves in figure 5.1. For a given crack length we can estimate the gradient of each curve and note the corresponding ΔK value. If we now plot $log(da/dN)$ against $log(\Delta K)$ we obtain the curve shown in figure 5.2 which may be considered to consist of three regions. In region

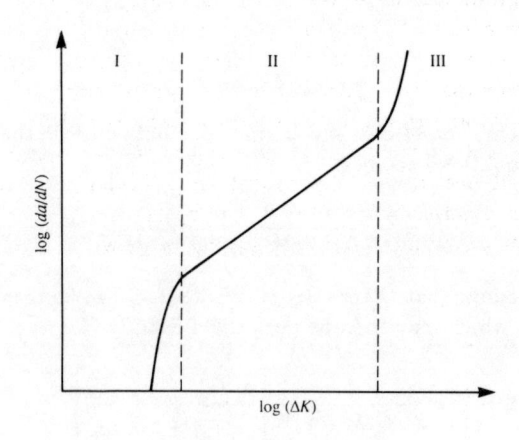

Figure 5.2 Logarithmic crack growth rate

I, the crack does not start to grow until ΔK reaches a critical threshold value ΔK_0 regardless of how many cycles are applied. However, as soon as a crack begins to grow, the crack growth rate increases very rapidly as ΔK increases with crack length. The area of most interest is region II, for this contains cracks of detectable size which grow in such a way that the log-log graph is almost linear. In region III, the crack is large and plasticity begins to dominate elasticity as the crack grows to failure.

In the analysis of fatigue crack growth it is usual to restrict attention to region II and to assume a linear relationship (i.e. an equation of the form $y = mx + c$ between $log(da/dN)$ and $log(\Delta K)$ so that:

$$log(da/dN) = m log(\Delta K) + log(C) \tag{5.3}$$

Taking antilogs of the above equation leads to the *Paris equation*

$$da/dN = C(\Delta K)^m \tag{5.4}$$

where m and C are empirical constants for a given material. m is often called the *Paris exponent* and is typically in the range 3-4 for common steels and aluminium alloys. The determination of a Paris-type law is considered as an example in chapter 6, section 6.

We can now use the Paris law to predict the number of cycles required for a crack to grow from length a_I to a_F. Rearranging equation (5.4) and integrating:

$$N = \int_{a_I}^{a_F} \frac{1}{C(\Delta K)^m} da \tag{5.5}$$

Using the definition of ΔK and taking constants outside the integral, the above equation becomes:

$$N = \frac{1}{C(\Delta\sigma\sqrt{\pi})^m} \int_{a_I}^{a_F} \frac{1}{Q^m a^{m/2}} da \tag{5.6}$$

Now if we assume that Q does not vary with a, Q can be taken outside the integral which may then be performed exactly (for $m \neq 2$) giving

$$N = \frac{1}{C(Q\Delta\sigma\sqrt{\pi})^m} \left[\frac{a_I^{1-m/2} - a_F^{1-m/2}}{m/2 - 1} \right] \tag{5.7}$$

If Q is a known function of a (fitting a polynomial to data if necessary) then it may be possible to integrate exactly. It is generally more convenient to integrate numerically. Program 5.2 below uses Simpson's rule (section 3.3) to integrate equation (5.6) and hence obtain the required number of cycles for a given increase in crack length. As an example we first take a central crack in a large sheet so that $Q = 1$.

Program 5.2 PCG: Crack growth using the Paris law.

```
100 PRINT " FATIGUE CRACK GROWTH USING THE PARIS LAW "
110 INPUT " STRESS RANGE (MPA)";R
120 INPUT " INITIAL CRACK LENGTH AI (M)";A0
```

Fatigue crack growth

```
130 INPUT " FINAL CRACK LENGTH AF (M)";A2
140 INPUT " PARIS EXPONENT ";M
150 INPUT " PARIS CONSTANT C ";C
160 PRINT
170 P1=4*ATN(1.0)
180 INPUT "  EVEN NUMBER OF INTERVALS (0 TO STOP)";N
190 IF N=0 GOTO 380
200 H=(A2-A0)/N
210 DEF FNQ(A)=1
220 REM INTEGRAL OF 1/(Q SQRT(A) )**M USING SIMPSONS RULE
230 T=0
240 S2=1/(SQRT(A0)*FNQ(A0))**M
250 FOR I=0 TO (N-2) STEP 2
260 S0=S2
270 A1=A0+(I+1)*H
280 S1=1/(SQRT(A1)*FNQ(A1))**M
290 A2=A0+(I+2)*H
300 S2=1/(SQRT(A2)*FNQ(A2))**M
310 T=T+S0+4*S1+S2
320 NEXT I
330 T=T*H/3
340 REM FIND NUMBER OF CYCLES
350 T=T/(C*(R*SQRT(P1))**M)
360 PRINT " REQUIRED NUMBER OF CYCLES =";T
370 GOTO 160
380 END
```

Sample run 1

```
RUN PCG1

 FATIGUE CRACK GROWTH USING THE PARIS LAW
 STRESS RANGE (MPA)? 80
 INITIAL CRACK LENGTH AI (M)? .005
 FINAL CRACK LENGTH AF (M)? .01
 PARIS EXPONENT ? 3
 PARIS CONSTANT C ? 4.5E-11

  EVEN NUMBER OF INTERVALS (0 TO STOP) 4
 REQUIRED NUMBER OF CYCLES = 64598
```

```
  EVEN NUMBER OF INTERVALS (0 TO STOP) 8
  REQUIRED NUMBER OF CYCLES = 64574.2

  EVEN NUMBER OF INTERVALS (0 TO STOP) 0
Ready
```

Program notes

(1) The function $Q(a)$ is defined in line 210. Because this function is a constant the result may be checked with the exact value from equation (5.7), that is 64572.4 cycles.

(2) The integration is repeated to check that enough intervals are being used. Since there is little change in the result, this is clearly the case.

Sample run 2

```
RUN PCG2

  FATIGUE CRACK GROWTH USING THE PARIS LAW
  STRESS RANGE (MPA)? 80
  INITIAL CRACK LENGTH AI (M)? .005
  FINAL CRACK LENGTH AF (M)? .01
  PARIS EXPONENT ? 3
  PARIS CONSTANT C ? 4.5E-11

  EVEN NUMBER OF INTERVALS (0 TO STOP) 4
  REQUIRED NUMBER OF CYCLES = 46113.2

  EVEN NUMBER OF INTERVALS (0 TO STOP) 8
  REQUIRED NUMBER OF CYCLES = 46096.3

  EVEN NUMBER OF INTERVALS (0 TO STOP) 0
Ready
```

Program notes

(1) The function $Q(a)$ is defined in line 210 and is the polynomial for an edge crack in a long strip of width W given by Rooke and Cartwright (1976). Line 205 has been added to define W. The modifications to PGC1 are:

```
205 W=1
210 DEF FNQ(A)=1.12-0.23*A/W+10.6*(A/W)**2
   -21.7*(A/W)**3+30.4*(A/W)**4
```

(2) Although the crack is very small relative to the strip width, the stress intensity factor is significantly higher than for the first example. As a result, the number of cycles for the same growth is reduced from 64,000 to 46,000.

5.2.1 R-values

So far in this chapter we have assumed that crack growth depends only on the change in stress $\Delta\sigma$. In fact, the overall level of stress also has an effect. If, for example, a specimen has a yield stress of 180 MPa and a cyclic stress of 140-160 MPa is applied, then one might expect this to give a higher growth rate than a similar cycle from 20-40 MPa. One attempt to characterise this problem is to specify an R-value where

$$R = K_{min}/K_{max} \tag{5.8}$$

High R-values correspond to relatively high average stress levels and may give higher crack growth rates. Low R-values correspond to relatively low average stress levels and may give lower growth rates. The effect of R seems to be rather limited for region II crack growth in high strength steels but it may still have a considerable effect on near-threshold growth. R-values seem to be important in all regions for aluminium alloys. When selecting values for use in a Paris-type law it is wise to ensure that the values chosen are derived from a test with a similar R-value to the problem under consideration. Another difficulty that can arise is that the stress may oscillate between a positive and a negative value. The simplest analysis indicates that the crack surfaces will close and the SIF will become zero as soon as the stress becomes negative. The R-value is therefore zero. However, experiments with aluminium alloys show that this analysis is rather too simple and that negative R-values can reduce the crack growth rate to below that for zero R. Again, the conclusion is that experimental studies of crack growth rates should be consulted before choosing parameters. A sketch graph showing variation in growth rate with R-value for an aluminium alloy is given in figure 5.3

A hostile chemical environment may also increases crack growth rates via stress corrosion. An atmosphere as apparently innocuous as

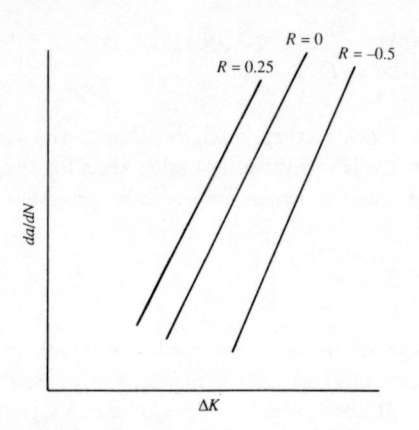

Figure 5.3 Typical effects of R in aluminium alloys

moist air can increase growth rates by as much as 50 percent in some alloy steels. Effects governing crack growth rates are introduced more fully by Ewalds and Wanhill (1985).

5.2.2 Variable amplitude loading

The assumption that the stress range $\Delta\sigma$ is a constant does not hold for a considerable number of practical problems. Suppose that an otherwise constant amplitude cyclic loading contains a single large stress cycle or overload (figure 5.4). This large stress creates a large plastic zone ahead of the crack tip which is subject to the usual residual compressive stresses. The relatively small stress cycles following the overload cause the crack to grow more slowly than expected because of the large residual stresses. As the crack grows through the plastic zone, its rate of growth increases until the tip reaches a point where the zone size corresponds to that for the current stress amplitude. The crack then grows at the constant amplitude rate. In figure 5.4, constant and variable amplitude crack growth is illustrated. Note that the effect of a large positive stress is much greater than that produced by a large positive-negative stress cycle.

If the stress cycles are of randomly varying heights then a root mean square average of ΔK may be used in the Paris law. If, however, the stress cycles form a repeated pattern, then a close analysis is necessary which lies beyond the scope of this book.

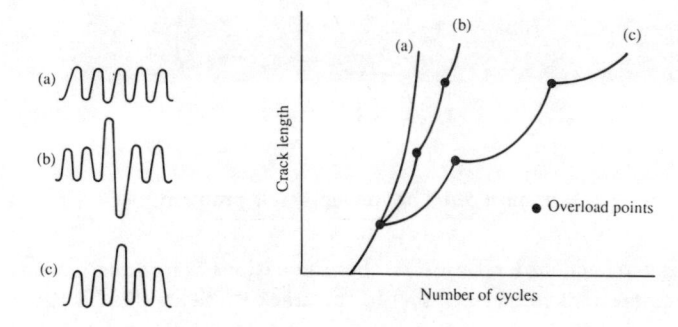

Figure 5.4 Variable amplitude crack growth

5.3 Second order terms

In discussing fatigue crack growth it was assumed that the crack would always grow along the crack line, implicitly assuming that K_I is the only stress intensity factor. Real cracks are not always as obliging ! If both K_I and K_{II} (or K_{III}) are non-zero then a crack is said to be under mixed-mode loading. Suprisingly, there is not yet a universally accepted way of predicting crack growth under this type of loading. Two parameters are critical: the direction in which the crack will grow and the effective stress intensity factor K_e in that direction. These parameters appear to be influenced significantly by the second order terms in the crack tip expansions of chapter 2 and so we first consider the effect of these additional terms on stresses and displacements.

To illustrate the effect of second order terms, we will consider a two-dimensional sheet containing a crack at an angle α to the applied stress which is shown in figure 5.5. If the sheet is infinite, then the analytical solution is well-known. This is a mixed-mode problem if α is not a right angle and hence both K_I and K_{II} will be non-zero. Equations for the stresses and displacements close to the crack tip were defined in chapter 2 in terms of the first order constants K_I and K_{II}.

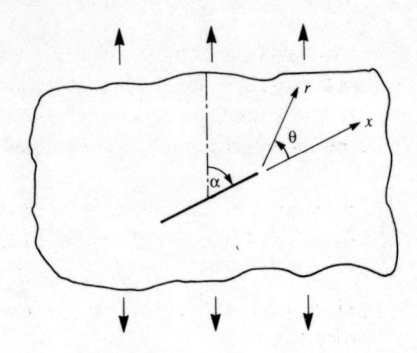

Figure 5.5 The angled crack problem

Adding second order terms to those equations we obtain improved approximations (superscript ∗) to the crack tip fields (Smith 1987):

$$\begin{aligned}
\sigma_x^* &= \sigma_x + L_I \\
\sigma_y^* &= \sigma_y \\
\tau_{xy}^* &= \tau_{xy} \\
u_x^* &= u_x + \frac{L_I}{8\mu}(1+\kappa)(r\cos\theta + a) - \frac{L_{II}}{8\mu}(1+\kappa)r\sin\theta \\
u_y^* &= u_y + \frac{L_{II}}{8\mu}(1+\kappa)(r\cos\theta + a) - \frac{L_I}{8\mu}(3-\kappa)r\sin\theta
\end{aligned} \quad (5.9)$$

For the infinite plate case, all the constants are known from the analytical solution so that

$$\begin{aligned}
K_I &= \sigma\sqrt{\pi a}\sin^2\alpha \\
K_{II} &= \sigma\sqrt{\pi a}\sin\alpha\cos\alpha \\
L_I &= \sigma\cos 2\alpha \\
L_{II} &= \sigma\sin 2\alpha
\end{aligned} \quad (5.10)$$

The only effect of second order terms on the stresses is an additional constant stress L_I parallel to the crack with normal and shear stresses unchanged. The displacements are changed quite significantly in both directions, although the term multiplying the crack length a does not depend on θ and is thus effectively constant.

Second order terms

It is very important to appreciate that the elastic material is defined to lie between the lines $\theta = -180°$ and $\theta = +180°$ (which are separated by the crack surfaces). Only angles in this range should be used or the values obtained may be incorrect, as for any formulae based on crack tip expansions.

The program CRSS given in chapter 2 may be modified to include these second order terms by adding the following lines:

Program Modification 5.3 CRSSECO: Crack stresses and displacements with second order terms

```
145 INPUT " LI      LII ",L1,L2
147 INPUT " CRACK LENGTH A";AC

315 C1=COS(T)
316 S1=SIN(T)

375 E1=(1.0+K)/(8*M)
376 E2=(3-K)/(8*M)

404 U1=U1+L1*E1*(R*C1+AC)-L2*E1*R*S1
406 U2=U2+L2*E1*(R*C1+AC)-L1*E2*R*S1
408 PRINT "SECOND ORDER                    UX =";U1," UY =";U2

465 PRINT "SECOND ORDER SX =";T1+L1
```

Sample run 1

```
RUN CRSSECO

DISPLACEMENTS AND STRESSES AT DISTANCE (R),
ANGLE (THETA) FROM A CRACK TIP
TERMINATES IF (R) IS NEGATIVE. UNITS MN, METRES

   KI    KII  ? 100,0
   LI    LII  ? -56.4,0
   CRACK LENGTH A? .1
   PLANE STRESS (Y/N)        ? Y
```

```
DISTANCE R, ANGLE THETA (DEGREES)? .0001,180
PLANE STRESS     UX =-.194204E-12        UY = .797885E-05
SECOND ORDER     UX =-.281718E-04        UY = .797885E-05
    SX =-.348766E-03     SY = 0      TXY =-.20795E-11
SECOND ORDER SX =-56.4003

DISTANCE R, ANGLE THETA (DEGREES)? .0001,-180
PLANE STRESS     UX =-.194204E-12        UY =-.797885E-05
SECOND ORDER     UX =-.281718E-04        UY =-.797885E-05
    SX =-.348766E-03     SY = 0      TXY = .20795E-11
SECOND ORDER SX =-56.4003

DISTANCE R, ANGLE THETA (DEGREES)? .0001,0
PLANE STRESS     UX = .27926E-05         UY = 0
SECOND ORDER     UX =-.254356E-04        UY = 0
    SX = 3989.42         SY = 3989.42    TXY = 0
SECOND ORDER SX = 3933.02

DISTANCE R, ANGLE THETA (DEGREES)? -1,0

Ready
```

Program notes

(1) This first run shows that second order terms may have an effect even for mode I loading. There is just a relatively small additional stress in the x-direction but a substantial change in the x-displacements due mainly to the crack length a. However, the crack sliding displacement is not affected since there is no difference between the movements of the crack surfaces ($\theta = \pm 180°$).

Sample run 2

```
RUN CRSSECO

DISPLACEMENTS AND STRESSES AT DISTANCE (R),
ANGLE (THETA) FROM A CRACK TIP
TERMINATES IF (R) IS NEGATIVE. UNITS MN, METRES

KI    KII    ? 44.3,76.7
LI    LII    ? 50,86.6
CRACK LENGTH A? .01
```

```
PLANE STRESS (Y/N)           ? N
DISTANCE R, ANGLE THETA (DEGREES)? .0001,180
PLANE STRAIN    UX = .5569E-05         UY = .321651E-05
SECOND ORDER    UX = .782125E-05       UY = .711741E-05
   SX =-6119.77      SY =-.159498E-11    TXY =-.267504E-03
SECOND ORDER SX =-6069.77

DISTANCE R, ANGLE THETA (DEGREES)? .0001,-180
PLANE STRAIN    UX =-.5569E-05         UY =-.321651E-05
SECOND ORDER    UX =-.331675E-05       UY = .684385E-06
   SX = 6119.77      SY = .159498E-11    TXY =-.267504E-03
SECOND ORDER SX = 6169.77

DISTANCE R, ANGLE THETA (DEGREES)? .0001,0
PLANE STRAIN    UX = .919003E-06       UY =-.159114E-05
SECOND ORDER    UX = .321675E-05       UY = .238856E-05
   SX = 1767.31      SY = 1767.31        TXY = 3059.89
SECOND ORDER SX = 1817.31

DISTANCE R, ANGLE THETA (DEGREES)? 0,-1
```

Program notes

(1) This example is for a true mixed-mode case ($\alpha = 30°$). Plane strain conditions result in slightly different displacements to the plane stress case but a much smaller crack length reduces the effect of additional constants.

(2) The crack opening and sliding displacements are the same for both first and second order terms. Check this by subtracting the displacements for $\theta = -180°$ from those for $\theta = 180°$.

(3) When $\theta = 0°$ there is a large shear stress near the crack tip due to the mode II loading.

5.4 Predicting the direction of crack growth

A number of criteria have been proposed which aim to predict the direction in which a crack will propagate under mixed-mode loading, none of which has proved entirely satisfactory. We will consider two common criteria here and also the effect of second order terms on these criteria.

The *maximum tensile stress* (MTS) criterion assumes that a crack will grow in the direction θ for which the circumferential tensile stress $t_{\theta\theta}$ is a maximum. An expression for tensile stress may be derived from equations (5.9) (by using the stress transformation formulae in chapter 2) as:

$$t_{\theta\theta} = \frac{1}{4\sqrt{2\pi r}}[K_I(3\cos\theta/2 + \cos 3\theta/2) - 3K_{II}(\sin\theta/2 + \sin 3\theta/2)]$$
$$+ \; L_I \sin^2\theta \qquad (5.11)$$

Assuming that only first order terms are to be used, then, to find the maximum $t_{\theta\theta}$ the above expression (with $L_I = 0$) is differentiated with respect to θ and set to zero. After some manipulation using basic trigonometry formulae, the predicted growth direction θ_0 is given by:

$$K_I \sin\theta_0 + K_{II}(3\cos\theta_0 - 1) = 0 \qquad (5.12)$$

This equation may now be solved for θ_0 by trial and error or by a non-linear equation method such as Newton-Raphson.

Now when second order terms are included it is no longer possible to eliminate r from equation (5.11) since it is no longer a common factor. The circumferential stress is therefore a function of the distance r from the crack tip and hence so is the predicted direction of crack growth. The optimum value of r may be material dependent and related to the size of the plastic zone, here we will take $r/a = 0.01$ to illustrate second order effects. It is not straightforward to differentiate and find a maximum with two terms; the simplest approach is to evaluate $t_{\theta\theta}$ for a range of angles and estimate the maximum by trial and error.

The *strain energy density* criterion predicts that crack growth occurs in the direction in which strain energy density (SED) is a minimum (Sih 1973). This direction corresponds to that of maximum potential energy density. In two dimensions the strain energy density is

$$\frac{dW}{dA} = \frac{1}{2E}(\sigma_x^2 + \sigma_y^2) - \frac{\nu}{E}\sigma_x\sigma_y + \frac{1}{2\mu}\tau_{xy}^2 \qquad (5.13)$$

Now if the stresses are represented by two term expansions then equation (5.9) can be used to substitute into the above equation and after considerable manipulation:

$$\frac{dW}{dA} = \frac{K_I^2}{16\mu\pi r}[(\kappa - \cos\theta)(1 + \cos\theta)]$$

$$+ \ \frac{K_I K_{II}}{8\mu\pi r} \sin\theta \left[(1-\kappa) + 2\cos\theta\right]$$

$$+ \ \frac{K_{II}^2}{16\mu\pi r} \left[(\kappa - 3\cos\theta)(1-\cos\theta) + 4\cos\theta\right]$$

$$+ \ \frac{K_I L_I \cos\theta/2}{4\mu\sqrt{2\pi r}} \left[\kappa - \cos\theta - 2\sin^2\theta\right]$$

$$- \ \frac{K_{II} L_I \sin\theta/2}{4\mu\sqrt{2\pi r}} \left[\kappa + \cos\theta + 2\cos^2\theta\right] + \frac{(1+\kappa)}{16\mu} L_I^2 \quad (5.14)$$

This SED function is rather complicated and may, in fact, possess several minima. The highest mimimum should be chosen as the crack growth direction. Note that the use of second order terms for SED may mean that occasionally this function does not possess a minimum. This problem has not yet been satisfactorily resolved.

In the simplest possible case when the crack is under pure mode I loading, we would expect it to grow in the same plane ($\theta = 0°$). The strain energy density function has two minima, at $\theta = 0°$ and at $\theta = 180°$. The higher of the two minima is at $\theta = 0°$ and this is therefore the predicted direction of growth. The maximum tensile stress also occurs at $\theta = 0°$. Although the SED and the MTS may give different values in their first and second order approximations, the corresponding minima and maxima are at the same angle for this elementary case.

For more difficult examples a computer program is useful and a program which calculates the tensile stress and the strain energy density is given below. The user may select the starting angle and the change in angle step for 12 steps. The sequence can be repeated for smaller angular steps without re-inputting initial values to 'home in' on the desired turning point. The number of steps and the material constants are included in the program for simplicity, though this would be easy to change if required.

Program 5.4 SED: Strain energy density and tensile stress with second order terms

```
100 PRINT "STRAIN ENERGY DENSITY AND STRESS AT DISTANCE (R),"
110 PRINT "ANGLE (THETA) FROM A CRACK TIP"
120 PRINT "TERMINATES IF STEP LENGTH -VE, UNITS MN, METRES"
130 PRINT
140 INPUT " KI     KII ";K1;K2
```

```
150 INPUT " LI       ";L1
160 INPUT " DISTANCE R FROM CRACK TIP";R
170 REM P1=PI, E=YOUNGS MODULUS, N=POISSONS RATIO (NU)
180 P1=4.0*ATN(1.0)
190 E=200000
200 N=0.3
210 REM M=SHEAR MODULUS (MU), K=KAPPA FOR PLANE STRESS
220 M=E/(2*(1+N))
230 K=(3-N)/(1+N)
240 T$="PLANE STRESS"
250 INPUT " PLANE STRESS (Y/N)",A$
260 IF(A$="Y") GOTO 300
270 T$="PLANE STRAIN"
280 REM  KAPPA FOR PLANE STRAIN
290 K=3-4*N
300 PRINT
310 INPUT " START ANGLE (DEG), STEP LENGTH (DEG) ";A,H
320 IF (H<0) GOTO 600
330 N1=12
340 PRINT "ANGLE  SED ONE   SED TWO   TSTRESS(1T)  TSTRESS(2T)"
350 B1=1/(16*M*P1*R)
360 B2=1/(4*M*SQRT(2*P1*R))
370 REM *** FIND SED AND TENSILE STRESS FOR N1 ANGLES
380 FOR I=0 TO N1
390 REM  CONVERT ANGLE TO RADIANS
400 A1=(A+I*H)*P1/180
410 C=COS(A1)
420 S=SIN(A1)
430 REM EVALUATE STRAIN ENERGY DENSITY
440 T1=B1*K1*K1*(K-C)*(1+C)
450 T2=2*B1*K1*K2*S*((1-K)+2*C)
460 T3=B1*K2*K2*((1+K)*(1-C)+(1+C)*(3*C-1))
470 T4=B2*K1*L1*COS(A1/2)*(K-2.0-C+2*C*C)
480 T5=B2*K2*L1*SIN(A1/2)*(K+C+2*C*C)
490 T6=L1*L1*(1+K)/(16*M)
500 D=T1+T2+T3
510 D2=D+T4-T5+T6
520 REM EVALUATE TENSILE STRESS T(@@)
530 F1=3.0*COS(A1/2.0)+COS(1.5*A1)
540 F2=SIN(A1/2.0)+SIN(1.5*A1)
550 T1=B2*M*(K1*F1-3*K2*F2)
560 T2=T1+L1*S*S
570 PRINT A1*180/P1,D,D2,T1,T2
```

```
580 NEXT I
590 GOTO 300
600 END
```

Sample run 1

```
RUN SED

STRAIN ENERGY DENSITY AND STRESS AT DISTANCE (R),
ANGLE (THETA) FROM A CRACK TIP
TERMINATES IF STEP LENGTH NEGATIVE, UNITS MN, METRES

  KI    KII ? 0,1
  LI      ? 0
  DISTANCE R FROM CRACK TIP? .01
  PLANE STRESS (Y/N)        ? N

  START ANGLE (DEG), STEP LENGTH (DEG) ? -180,30
    ANGLE    SED ONE      SED TWO     TSTRESS(1T)   TSTRESS (2T)
    -180    .144831E-03  .144831E-03    0              0
    -150    .122662E-03  .122662E-03   .774403        .774403
    -120    .762949E-04  .762949E-04   2.59121        2.59121
    -90     .465528E-04  .465528E-04   4.23142        4.23142
    -60     .556048E-04  .556048E-04   4.4881         4.4881
    -30     .868257E-04  .868257E-04   2.89012        2.89012
     0      .103451E-03  .103451E-03    0              0
     30     .868257E-04  .868257E-04  -2.89012       -2.89012
     60     .556048E-04  .556048E-04  -4.4881        -4.4881
     90     .465528E-04  .465528E-04  -4.23142       -4.23142
    120     .762949E-04  .762949E-04  -2.59121       -2.59121
    150     .122662E-03  .122662E-03  -.774403       -.774403
    180     .144831E-03  .144831E-03    0              0

  START ANGLE (DEG), STEP LENGTH (DEG) ? -90,5
    ANGLE    SED ONE      SED TWO     TSTRESS(1T)   TSTRESS (2T)
    -90     .465528E-04  .465528E-04   4.23142        4.23142
    -85     .453389E-04  .453389E-04   4.39518        4.39518
    -80     .452996E-04  .452996E-04   4.51447        4.51447
    -75     .463952E-04  .463952E-04   4.58576        4.58576
    -70     .485525E-04  .485525E-04   4.60629        4.60629
    -65     .516665E-04  .516665E-04   4.57411        4.57411
    -60     .556048E-04  .556048E-04   4.4881         4.4881
    -55     .602111E-04  .602111E-04   4.34805        4.34805
```

-50	.65311E-04	.65311E-04	4.15462	4.15462
-45	.707167E-04	.707167E-04	3.90932	3.90932
-40	.762338E-04	.762338E-04	3.61455	3.61455
-35	.816668E-04	.816668E-04	3.2735	3.2735
-30	.868257E-04	.868257E-04	2.89012	2.89012

```
     START ANGLE (DEG), STEP LENGTH (DEG) ? -75,1
    ANGLE     SED ONE       SED TWO      TSTRESS(1T)   TSTRESS (2T)
```

ANGLE	SED ONE	SED TWO	TSTRESS(1T)	TSTRESS (2T)
-75	.463952E-04	.463952E-04	4.58576	4.58576
-74	.467447E-04	.467447E-04	4.59401	4.59401
-73	.471359E-04	.471359E-04	4.6002	4.6002
-72	.475682E-04	.475682E-04	4.60432	4.60432
-71	.480407E-04	.480407E-04	4.60636	4.60636
-70	.485525E-04	.485525E-04	4.60629	4.60629
-69	.491026E-04	.491026E-04	4.60412	4.60412
-68	.496901E-04	.496901E-04	4.59983	4.59983
-67	.50314E-04	.50314E-04	4.5934	4.5934
-66	.509731E-04	.509731E-04	4.58483	4.58483
-65	.516665E-04	.516665E-04	4.57411	4.57411
-64	.523929E-04	.523929E-04	4.56123	4.56123
-63	.531512E-04	.531512E-04	4.54619	4.54619

```
     START ANGLE (DEG), STEP LENGTH (DEG) ? 0,-1

Ready
```

Program notes

(1) We start with the case of a crack under pure shear. A more careful study shows that SED has two minima which are identical in size while the MTS has a maximum at between $-90°$ and $-60°$. The fact that there are two identical SED minima should not be confusing since we should have an intuitive idea of the correct direction and also we know the direction of maximum MTS.

(2) The appropriate SED minimum or MTS maximum is selected and the loop repeated with a smaller step. The step size is reduced again for a more accurate result.

(3) We assume $L_I = 0$ and therefore the first and second order approximations are identical.

(4) A change in the value of Poisson's ratio does not affect the MTS criterion at all and the predicted angle is constant at $\theta = -70.5°$.

However the angle predicted by the SED criterion ranges from $\theta = -70.5°$ at $\nu = 0$ to $\theta = -90°$ at $\nu = 0.5$.

Sample run 2

```
RUN SED

STRAIN ENERGY DENSITY AND STRESS AT DISTANCE (R),
ANGLE (THETA) FROM A CRACK TIP
TERMINATES IF STEP LENGTH -VE, UNITS MN, METRES


    KI      KII ? 44.3,76.7
    LI       ? 50
    DISTANCE R FROM CRACK TIP? .01
    PLANE STRESS (Y/N)           ? N

START ANGLE (DEG), STEP LENGTH (DEG) ? -180,30
 ANGLE    SED ONE      SED TWO      TSTRESS(1T)    TSTRESS (2T)
 -180     .852025      .996937      -.526701E-05   -.526701E-05
 -150     .962244      1.10093       62.4608        74.9608
 -120     .781175      .875861       220.837        258.337
  -90     .505827      .57074        387.034        437.034
  -60     .395648      .465974       459.028        496.528
  -30     .517338      .588678       380.946        393.446
    0     .689798      .71846        176.731        176.731
   30     .681148      .645261      -62.3977       -49.8977
   60     .456531      .387632      -229.447       -191.947
   90     .224622      .162961      -262.066       -212.066
  120     .233231      .172896      -176.654       -139.154
  150     .517229      .422115      -56.3326       -43.8326
  180     .852025      .718488      -.526701E-05   -.526701E-05

START ANGLE (DEG), STEP LENGTH (DEG) ? -65,1
 ANGLE    SED ONE      SED TWO      TSTRESS(1T)    TSTRESS (2T)
  -65     .396197      .46438        456.857        497.927
  -64     .395505      .464112       457.636        498.027
  -63     .395105      .464141       458.243        497.937
  -62     .394998      .464464       458.678        497.658
  -61     .395179      .465077       458.94         497.188
  -60     .395648      .465974       459.028        496.528
  -59     .396401      .46715        458.941        495.678
  -58     .397435      .4686         458.679        494.638
  -57     .398747      .470318       458.241        493.409
```

-56	.400333	.472298	457.626	491.991
-55	.402188	.474534	456.835	490.385
-54	.404307	.477017	455.866	488.592
-53	.406687	.479742	454.721	486.612

START ANGLE (DEG), STEP LENGTH (DEG) ? 0,-1

Program notes

(1) Taking $\alpha = 30°$ in the inclined crack problem gives the stress intensity factor data used for input.

(2) The strain energy density has two minima, the highest is around -60 degrees so we examine this region more closely.

(3) The estimates of crack growth direction are now all different. A closer examination shows that SED one term gives $-62.1°$, SED two term gives $-63.6°$, MTS one term gives $-60.0°$ and MTS two term gives $-64.0°$. The range of predicted crack paths is shown in figure 5.6.

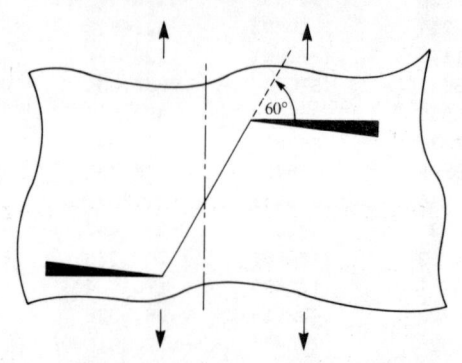

Figure 5.6 Direction of crack growth

The overall variation in strain energy density and tensile stress for the previous example is shown in figure 5.7. Note that that none of the parameters are symmetric around $\theta = 0°$ for this mixed-mode case. Which of these predictors is the better ? Unfortunately, because of the scatter in experimental measurements it is not possible to give a precise answer. However a suitable choice of distance r can make the two term MTS criterion agree best with some experimental data (Smith 1987). The optimal value of r is likely to be material dependent and more consistent experimental studies are required before a 'best'

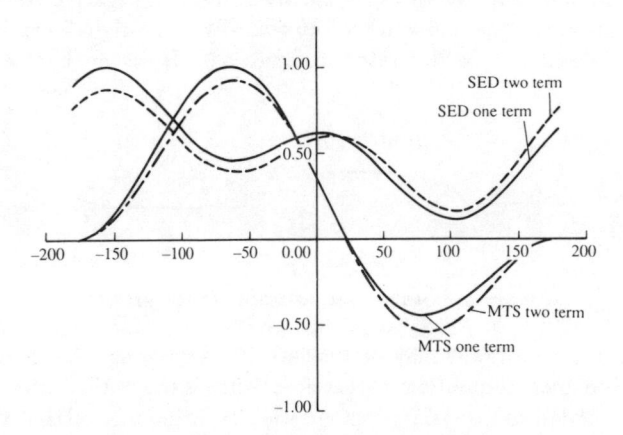

Figure 5.7 The variation of strain energy and tensile stress

model of crack growth direction can be selected. Remember that crack growth into different stress fields or in a new direction will usually change the stress intensity factors and hence the direction of crack growth leading to a curved path.

5.5 Fast fracture under mixed-mode loading

In Chapter 4, the critical value of the stress intensity factor which resulted in sudden and uncontrolled crack growth - fast fracture - is called the fracture toughness K_{Ic}. Under the assumption of a constant mode I load in an infinite sheet, the fracture toughness is simply related to the critical stress σ_c by the equation:

$$K_{Ic} = \sigma_c\sqrt{\pi a} \tag{5.15}$$

From section 5.4, the initial direction of crack growth under pure mode I loading is expected to be in the crack plane and this is certainly true for fast fracture. The crack line may, of course, subsequently curve away to remain approximately normal to the maximum tensile stress (or minimum SED).

It can be shown that the fast fracture path of a crack is unstable and hence liable to change direction in response to any small imperfection if the second order stress L_I is positive. If L_I is negative, then the crack path is stable and will tend to return to the original crack line. This phenomenon is illustrated in figure 5.8. However, by the time

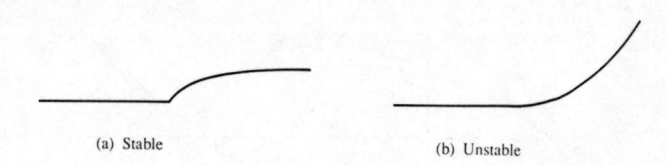

(a) Stable (b) Unstable

Figure 5.8 Stable and unstable crack growth

fast fracture occurs it may be too late to worry about the direction. The fundamental question is, therefore, what is the critical load under mixed-mode conditions ? This question is intimately related to the parameters discussed already, namely the maximum tensile stress and the strain energy density.

The maximum tensile stress criterion predicts that failure will occur when this stress reaches some critical value. If we only consider first order terms and take the coefficient of $1/\sqrt{r}$ in equation (5.11) as T, then

$$T = \frac{1}{4\sqrt{2\pi}}[K_I(3\cos\theta/2 + \cos 3\theta/2) - 3K_{II}(\sin\theta/2 + \sin 3\theta/2)] \tag{5.16}$$

Assuming mode I failure ($K_I = K_{Ic}$ and $K_{II} = 0$) with in-plane crack growth ($\theta = 0°$) gives the critical value of the tensile stress coefficient:

$$T_c = \frac{1}{\sqrt{2\pi}}K_{Ic} \tag{5.17}$$

The crack just grows when the maximum tensile stress reaches this value, so equating (5.16) and (5.17) gives the condition for failure as:

$$K_I(3\cos\theta/2 + \cos 3\theta/2) - 3K_{II}(\sin\theta/2 + \sin 3\theta/2) = 4K_{Ic} \tag{5.18}$$

In a similar way we can obtain an equation using the critical strain energy density S_c in the form:

$$a_{11}K_I^2 + a_{12}K_IK_{II} + a_{22}K_{II}^2 = S_c = \frac{K_{Ic}^2}{4\mu(1-2\nu)} \tag{5.19}$$

where the coefficients a_{11}, a_{12} and a_{22} correspond to those in equation (5.14). It is now possible to sketch a graph showing the limiting

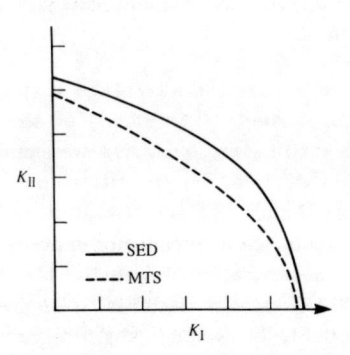

Figure 5.9 First order limiting values of K_I and K_{II}

regions defined by the two equations above and this has been done in figure 5.9. Experimental evidence suggests that the MTS criterion is a conservative one and it should therefore enclose a 'safe' region.

Equations (5.18) and (5.19) are obtained using only first order terms but they still require knowledge of the crack growth direction. Note that many older books on fracture mechanics give formulae which implicitly assume in-plane crack growth, although this may not be clearly stated. Since we require the direction of crack growth anyway, a simple procedure is as follows:

(1) Run Program 5.4 with $K_I = K_{Ic}$ and $K_{II} = L_I = 0$ and note the limiting values of strain energy density and maximum tensile stress at $\theta = 0$. Note that the critical MTS is not affected by second order terms since L_I is multiplied by $\sin\theta$ in equation (5.11) and $\theta = 0$. However, the critical SED is affected by second order terms and to compute these requires the value of σ_c as well as K_{Ic} from fracture toughness testing.

(2) Run Program 5.4 with all the appropriate values to find the angle of crack growth with your preferred criterion. If the value of SED or MTS at this angle is greater than the limiting value obtained in part (1) then assume failure will occur. Little data are available to help calculate

second order terms; the infinite sheet formulae (equation 5.10) can be used with a finite width correction if required but if this is not a satisfactory approximation, it may be necessary to use boundary or finite element methods.

The above procedure has the advantage that second order effects may be included if desired. The effects of second order terms on the critical fracture stress are even less well understood than their effects on fracture direction. Since critical parameters such as the distance at which SED and MTS are evaluated are likely to be material dependent, it is probably best to consider experimental evidence from each type of material separately and choose the criteria which seems most successful. If the same material is to be used again and again, then it would probably be worth using Program 5.4 to generate a graph like that shown in figure 5.9 to avoid repetitive calculations.

Example 5.5

Now suppose that we have a material with the properties included in program 5.4 and that its fracture toughness $K_{Ic} = 120 MPa\ m^{1/2}$. Running the program with $K_I = K_{Ic}$ and $K_{II} = L_I = 0$ gives the critical value of strain energy density as $0.596 MJ/m^2$, and that for maximum tensile stress as $479 MPa$. If K_I and K_{II} are the same as in sample run 2 of program 5.4, then the one term approximation gives a value for SED of $0.395 MJ/m^2$ at $-62.1°$ and a value for MTS of $459 MPa$ at $-60.0°$. Both the strain energy density and the maximium tensile stress are less than their corresponding critical values and hence predict no failure. It is interesting to note that the two-term maximium tensile stress approximation is more than the critical value of MTS and predicts failure. The one-term MTS criterion is usually conservative and so failure may not occur but unless there is strong experimental evidence to the contrary it would be unwise to allow the material to be loaded to the above level in service.

5.6 Fatigue under mixed-mode loading

As for fast fracture, there seems to be no one definitive theory which fits better than any other to a wide range of experiments. It does seem that initial direction can be predicted approximately by the strain energy density and maximum tensile stress criteria. However, instead of using the absolute K-values in equations (5.11) and (5.14), the stress intensity factor range ΔK should be used. The above two criteria are shown by Tanaka (1974) to agree fairly well with experimental growth directions when ΔK is close to the threshold value which causes crack growth. However he also observed that larger ΔK values (1.6 x threshold) gave predominantly plane stress conditions and crack growth then tended to be in line with the initial crack.

A Paris-type crack propagation law for mixed-mode cyclic loading may be written as:

$$da/dN = C(\Delta K_e)^m \tag{5.20}$$

where ΔK_e is an effective stress intensity factor. A number of forms have been suggested for ΔK_e but Tanaka claimed the best agreement with experimental values from:

$$\Delta K_e = (\Delta K_I^4 + 8\Delta K_{II}^4)^{1/4} \tag{5.21}$$

for a growth law where m is expected to be close to 4.

The threshold ΔK_0, the value below which no fatigue growth occurs is reasonably well approximated by the one-term strain energy density formula (equation 5.19) with S_c obtained from the mode I threshold value and ΔK used for K. The one term maximum tensile stress formula (equation 5.16) provides a conservative estimate of the mixed-mode threshold.

Example 5.6

A central crack of length $20mm$ in a large sheet is subjected to a cyclic loading of between 0 and $9.6 MPa$ at an angle of 60° to the crack. How long will the crack take to grow to $30mm$ if the material satisfies a Paris-type law as above, with $C = 1.9 \times 10^{-11}$ and $m = 4.3$?

Solution: Using equations 5.10 for an inclined crack in an infinite sheet with angle $\alpha = 60°$ gives

$$\Delta K_I = 1.80, \Delta K_{II} = 1.04 \quad (MPa\ m^{1/2})$$

and hence using equation (5.21)

$$\Delta K_e = \left(1.80^4 + 8 \times 1.04^4\right)^{1/4} = 2.12$$

Now program 5.2 (PCG1) is set up to use $\Delta\sigma$ so we can either change the program to accept ΔK or simply divide it by $\sqrt{\pi a}$ giving $\Delta\sigma = 8.44 MPa$.

```
RUN PCG1
FATIGUE CRACK GROWTH USING THE PARIS LAW
STRESS RANGE (MPA)? 8.44
INITIAL CRACK LENGTH AI (M)? .02
FINAL CRACK LENGTH AF (M)? .03
PARIS EXPONENT ? 4.3
PARIS CONSTANT C ? 1.9E-11

 EVEN NUMBER OF INTERVALS (0 TO STOP)? 4
REQUIRED NUMBER OF CYCLES = .136015E+08

 EVEN NUMBER OF INTERVALS (0 TO STOP)? 8
REQUIRED NUMBER OF CYCLES = .136003E+08

 EVEN NUMBER OF INTERVALS (0 TO STOP)? 0
Ready
```

The number of intervals used for integration appears to be adequate and a life of 13.6 megacycles is predicted.

5.7 References

Ewalds, H.L. and Wanhill R.J.H. (1985) *Fracture mechanics*, Edward Arnold, London.

Rooke, D.P. and Cartwright, D.J.(1976) *A compendium of stress intensity factors*, HMSO, London.

Sih, G.C., (1973) A special theory of crack propagation. in *Methods of Analysis and Solutions of Crack Problems*, G.C. Sih (Ed.), xxii-xlv, Noordhoff.

Smith, R.N.L. (1987) Second order terms and strain energy density for the angled crack problem. *Engg. Frac. Mech*, **26**, pp 463-469.

Tanaka, K., (1974) Fatigue crack propogation from a crack inclined to the cyclic stress axis, *Engg. Frac. Mech.*, **6**, 493-507.

PROBLEMS

(5.1) A long rectangular plate of width 0.5 m is under tension on the ends and has an edge crack at right angles to the edge of the plate. If the tension oscillates between 50 and 140 MPa how many cycles are required for the crack to grow from 2mm to 4mm and from

4mm to 6mm ? Take the Paris constant to be 2.64×10^{-12} and the corresponding exponent to be 3.3.

(5.2) A large plate contains a hole of radius 0.5 m with an edge crack of length 5mm at right angles to the applied stress. If the crack has a stress intensity factor of 1.1 times the infinite plate value, and assuming that the SIF remains constant, how long will it take the crack to grow to 7mm under a load varying from 30 to 90 MPa ? The Paris exponent is 3.1 and the constant C is 5×10^{-12}. Generate an approximate formula for the SIF and compare the crack growth rate with that obtained using a constant K. Hint: Assume infinite plate results.

(5.3) A large plate contains a central, normally loaded crack of length 10mm. If a randomly varying cyclic load is applied and a typical sequence of 3 cycles is $\pm 20 MPa$ $\pm 30 MPa$ $\pm 20 MPa$, how long will the crack take to double in size ? Take $m = 3.7$ and $C = 2.5 \times 10^{-12}$

(5.4) If a crack of length 0.3m is loaded so that $K_I = 50$, $K_{II} = 50$, $L_I = 0$, $L_{II} = 100$, find how much the crack opens at its centre if $E = 200,000 MPa$ and $\nu = 0.3$

(5.5) Examine the effect of varying E and ν on the stresses and displacements ahead of a crack tip. How do second order terms change this response ?

(5.6) A crack in an infinite sheet is under a constant normal load. If a loading parallel to the crack is added, does this affect the crack opening ? Does the new load change the likelyhood of fracture ?

(5.7) Derive the equations (5.11) and (5.12) for the maximum tensile stress criterion.

(5.8) Write down an expression for the strain energy density under pure mode I loading and hence obtain an equation for the direction of crack growth.

(5.9) If a crack is loaded as in question (5.4) above, in which direction is it most likely to grow ?

(5.10) A crack is loaded so that $K_I = K_{II} = 30 MPa\ m^{1/2}$. If the fracture toughness $K_{Ic} = 60 MPa\ m^{1/2}$, will fracture occur under plane strain or plane stress conditions ? Assume material constants are $E = 210,000 MPa$ and $\nu = 0.333$.

(5.11) If a material has a fracture toughness $K_{Ic} = 70 MPa\ m^{1/2}$, with $E = 210,000 MPa$ and $\nu = 0.333$, use program 5.4 to prepare a diagram showing the region in which fast fracture should not occur under mixed- mode loading.

Chapter 6

Some applications in fracture mechanics

6.1 Introduction

Fracture mechanics is still a relatively new field and with continual development in materials and material types, new methods are always required which can model experimental behaviour. In this chapter we aim to describe both some applications of fracture mechanics and some methods developed for particular problems. It is impossible to cover all applications or methods but it is to be hoped that the samples here will give the flavour of practical problems and approaches. Useful examples are also given by Parker (1981) and by Kanninen and Popelar (1985).

Starting with a brief discussion of some design approaches which aim to avoid the worst effects of fracture, we also mention the problem of determining crack size. The key quantity for fracture analysis is the critical crack size causing total failure and this is discussed in 6.3 where a BASIC program is used to avoid the tedious hand iterations traditionally used. Residual stresses may arise in a structure by design, which should be beneficial or by accident, which may not be! A simple example illustrates the potential dangers of residual weld stresses.

The most important part of fracture analysis is relating theoretical parameters to practical results and in 6.5 we describe one attempt to simplify the study of fatigue crack growth from notches. One consequence of this semi-heuristic model is an explanation of the observation that fatigue cracks at very sharp notches may actually stop growing.

Two simple but general techniques, for finding the gradient of a curve and for fitting a least square straight line are described in the context of finding a fatigue growth law and the corresponding BASIC programs are given in section 6.6.

Another semi-heuristic method is the failure assessment diagram which defines 'safe' loading for material behaving both elastically and for increasing degrees of plasticity up to fully plastic collapse. The production of such diagrams is really beyond the scope of this book but a simple example of their use is included.

Finally, in 6.8 we describe the difficulties of three dimensional fracture mechanics and discuss how published tables can be used to obtain a full range of fracture parameters. A BASIC program giving stress intensity factors for a semi-elliptical crack is demonstrated.

6.2 Design strategies

There are three common design strategies which aim to take account of the effects of fatigue crack growth. These are the 'safe-life', 'damage-tolerant' and 'fail-safe', methods.

The 'safe-life' approach is a traditional method which does not use the principles of fracture mechanics. It starts from curves of stress versus number of cycles (S-N curves) for the material being used. The time to failure on these curves for the type of loading being considered is estimated allowing statistically for experimental variation. The structure (a bridge, for example) is then designed to last for a fixed lifetime without inspection. This approach clearly requires a large factor of safety and factors of 5, or more, are commonly used.

The more recent 'damage-tolerant' approach is based on the assumption that cracks of some estimated maximum size exist anywhere within the structure. If a structure is under fatigue loading then these cracks will grow in the most highly stressed areas. It may be that non-destructive testing methods can identify the size and position of such cracks so that component lifetimes may be estimated. Even if some cracks are identified, it must be assumed that cracks of the largest size which could remain undetected also exist. Thus if a particular technique can identify cracks of say, 4mm in length, but not smaller cracks, a crack of 4mm must be assumed to exist in the most critical position within the component. Notwithstanding the above, in some fields there is a code of practice (a damage-tolerant requirement) which insists that cracks of a certain length are assumed to exist by the designer.

In order to ensure that a component does not suffer failure due to fracture, the designer must use the principles of fracture mechanics to provide answers to the following questions:

(1) What is the largest crack which may not be detected by the inspection techniques used ? (or which must be assumed by a damage-tolerant specification)
(2) What is the critical crack length at which sudden failure will occur ?
(3) How long will it take for a crack to grow from (1) to (2) ?

With the answers to the above questions the designer should also be able to state the maximum inspection interval and, if inspection is more frequent, the maximum crack size that may be allowed to continue in service.

The 'fail-safe' approach is not concerned with methods of fracture prediction but aims to design a structure so that even in the event of total failure of one part, the remainder can continue to operate safely. Needless to say, this is not easy to achieve but one example occurs in aircraft construction. A thin lightweight skin is supported by stiffeners or ribs which are firmly bonded to it in some way (Figure 6.1). A crack in the skin may develop and grow but when it reaches a stiffener it can grow no further because the stiffener prevents the crack opening. The stiffeners also reduce the vibrations which may be responsible for some crack growth. Another version of the 'fail-safe' approach is

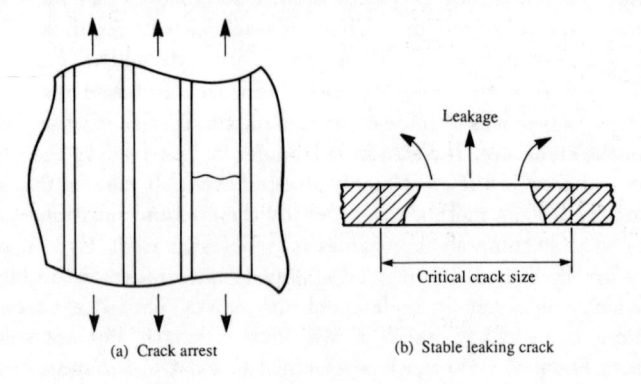

(a) Crack arrest   (b) Stable leaking crack

Figure 6.1 Fail-safe design

used in the design of pressure vessels where it is known as 'leak-before-break'. Since the total failure of a pressure vessel may be catastrophic, the vessels are designed so that the typical elliptic crack penetrates the skin of the vessel (Figure 6.1) allowing leakage to occur. Once the leakage is detected then preventative measures can be taken. The

Design Strategies

essential condition in 'leak-before-break' is that although the crack penetrates the vessel wall, it is still not wide enough to cause sudden failure of the whole vessel.

6.2.1 Crack Sizing

Techniques which allow components to be tested without loading to possible failure are known as non-destructive testing (NDT) techniques. Some common techniques are:

1) *Dye penetrant* - The surface of the component is coated with a penetrating fluid which is then wiped off. If the surface is then coated with a powder of contrasting colour, fluid retained in surface cracks discolours the powder.

2) *Magnetic* - Cracks in magnetic materials cause sharp changes in the local magnetic field. The surface is coated with a fluorescent liquid containing iron particles which show contrasting patterns around crack lines when the component is placed in a magnetic field.

3) *Radiography* - X-rays or Gamma rays can be used to produce a photographic plate on which the cracks show as darkened areas.

4) *Ultrasonic* - A pulse of high frequency sound waves is emitted by a tranducer on the material surface. By comparing the time delay of returning signals reflected from the far side of the material with signals reflected from cracks, the size and depth of the cracks may be estimated.

6.3 The critical crack length

The fracture toughness K_c is discussed in chapter 4 and is the value at which fast fracture occurs, given by:

$$K_c = Q\sigma_c\sqrt{\pi a_c} \qquad (6.1)$$

Hence the critical crack length a_c is

$$a_c = \left(\frac{K_c}{Q\sqrt{\pi}\sigma_c}\right) \qquad (6.2)$$

If Q is assumed to be constant then equation (6.2) may be used to find the critical crack length. If, as is more usual, Q varies with the crack length then an iterative procedure must be used to find a_c; first assuming a value for K, then computing a_c, then using a_c to find a new value of K and repeating the process until it converges. This is illustrated by the following example.

6.3.2 Example

An edge-cracked component has a nominal width $W = 250mm$, with $K_{Ic} = 70 MPa\, m^{1/2}$ and is loaded with an applied stress of $250 MPa$; What is the critical crack length at which the component will fail ? The only data available for the component have already been arranged in non-dimensional form and are given below.

$$a/W \qquad 0.04 \quad 0.1 \quad 0.2$$

$$Q(= K_I/\sqrt{\pi a}) \quad 1.128 \quad 1.2 \quad 1.48$$

Since K is not a constant, the first step is to obtain an expression for Q from the available data. This can be done using the Lagrange interpolation program from chapter 3.

```
RUN LAG

LAGRANGE INTERPOLATION
NUMBER OF POINTS TO BE GIVEN? 3
XI,YI? .04,1.128
XI,YI? .1,1.2
XI,YI? .2,1.48
COEFFICIENT OF X** 0    IS  1.12
COEFFICIENT OF X** 1    IS  -.199996
COEFFICIENT OF X** 2    IS  10
NEW X VALUE? .1
AT X = .1     Y = 1.2

ANOTHER X-VALUE (Y/N)? N
Ready
```

Q is therefore given by

$$Q = 1.12 - 0.2(a/W) + 10(a/W)^2 \tag{6.3}$$

and this expression can now be used in a program to determine the critical crack length iteratively using equation (6.2).

Design Strategies 125

Program 6.3 ACRIT: Critical crack length

```
100 PRINT " CRITICAL CRACK SIZE FROM KIC AND Q - (MN,METRES)"
110 INPUT " KIC =";K
120 INPUT " STRIP WIDTH W =";W
130 INPUT " MAXIMUM STRESS S =";S
140 P1=4.0*ATN(1.0)
150 C=S*SQRT(P1)
160 A=0.001
170 R=A/W
180 Q=1.12-0.2*R+10*R*R
190 A=(K/(Q*C))**2
200 PRINT " Q =";Q;" CRITICAL CRACK LENGTH A (METRES) =";A
210 PRINT
220 INPUT " REPEAT (Y/N) ";A$
230 IF A$="Y" GOTO 170
240 END
```

Sample run

```
RUN ACRIT

 CRITICAL CRACK SIZE FROM KIC AND Q - (MN,METRES)
 KIC =? 70
 STRIP WIDTH W =? .25
 MAXIMUM STRESS S =? 250
 Q = 1.11936  CRITICAL CRACK LENGTH A (METRES) = .199171E-01

 REPEAT (Y/N) ? Y
 Q = 1.16754  CRITICAL CRACK LENGTH A (METRES) = .183073E-01

 REPEAT (Y/N) ? Y
 Q = 1.15898  CRITICAL CRACK LENGTH A (METRES) = .185787E-01

 REPEAT (Y/N) ? Y
 Q = 1.16036  CRITICAL CRACK LENGTH A (METRES) = .185344E-01

 REPEAT (Y/N) ? N
Ready
```

Program notes

(1) The initial crack size is set to a small number in line 160 and the expression for Q in line 180 is then close to 1.12. Clearly the first value of Q should be non-zero.

(2) The program converges after 4 iterations to $a_c = 18.5mm$. If desired the relative size of the last change in a_c could be tested against some given error and the program stopped automatically.

(3) The program may not always converge very quickly but the user can help it along by changing the program to input an average value of A (at line 220 for example).

6.4 Residual stresses

Residual stresses are stresses which exist within a material in the absence of any external loads. They arise because of the way the material has been treated in its manufacture.

The process of shot peening, or bombarding the material surface with many small spherical particles, is intended to produce an outer layer of material with compressive residual stresses σ_r. Any crack which may exist in the material surface when it is loaded by a tensile stress σ is therefore affected by a net stress of $\sigma - \sigma_r$ since the principle of superposition allows us to add linear elastic terms. The crack will not now start to grow until the net stress exceeds the threshold value. Heat treatment may also be used to produce a similar effect to shot peening.

A word of caution is necessary however, since it is clear that the compressive stresses in the material surface must be produced by tensile stresses in a layer just below the surface. If the crack should penetrate the compressive surface layer then it will grow more rapidly than in untreated material until the crack tip grows through the tensile layer. It is clearly important to ensure that any compressive layer produced with the aim of preventing failure is sufficiently thick to completely inhibit crack growth.

Residual stresses are not always beneficial to the designer, for example an unfortunate side effect of the welding process may be the presence of tensile residual stresses σ_r which arise when the weld pool solidifies. Because of their origin these residual stresses may well be close to the yield strength of the material. A typical stress pattern around a weld is shown in figure 6.2. Under a tensile exterior loading

Residual stresses

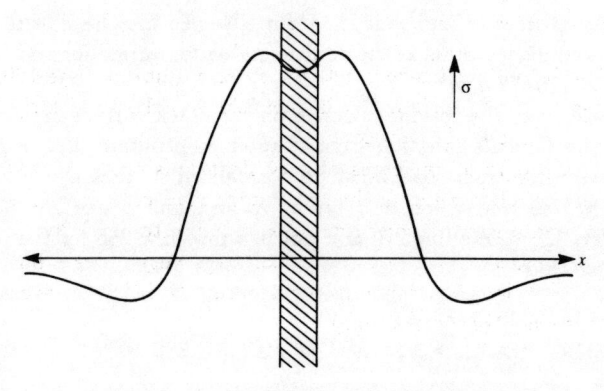

Figure 6.2 Typical weld stresses

σ the net stress will be $\sigma + \sigma_r$ which may be very high indeed! For this reason, highly stressed welded structures should be stress relieved by more uniform heating and cooling before use unless the design loading already takes account of residual stresses.

Example 6.4.1

A welded structure contains a crack of length 10 mm with a residual stress field as shown in figure 6.3. Assume the residual stress field has

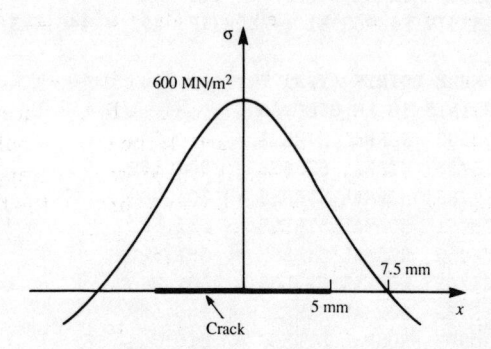

Figure 6.3 Crack in a residual stress field

the form
$$\sigma = 600 \cos\left(\frac{\pi x}{15}\right) \tag{6.4}$$

where x is in mm and σ is in MPa. Neglecting the effects of any other boundaries, what is the additional load required for fast fracture if $K_{Ic} = 110 MPa\ m^{1/2}$?

The effect of the residual stresses on the crack surface can be found using the Green's function from chapter 3, program 3.7. First, the expression for crack load must be normalised so that $a = 1$. We let $X = x/5$ so that when $x = 5mm$, $X = 1$ and hence $x = 5X$. In program 3.7, delete line 270 and put in a new line 255 for the residual stress so that

```
255 S=600*COS(P1*X/3)
260 PRINT " AT X = ";X*A;" NORMAL STRESS = ";S
```

Now we simply run the program to find the stress intensity factor; remember that with numerical integration a repeat run with more integration points helps to decide that enough points are being used.

```
RUN RES

SIF FOR A CRACK IN INFINITE SHEET.
 HALF CRACK LENGTH A ? 5
 NUMBER OF POINTS TO BE USED? 4
 AT X =  4.6194   NORMAL STRESS =   340.424
 AT X =  1.91342  NORMAL STRESS =   552.462
 AT X = -1.91342  NORMAL STRESS =   552.462
 AT X = -4.6194   NORMAL STRESS =   340.424
 STRESS INTENSITY FACTOR KI [/ SQRT(PI A)] = 446.443

 REPEAT FOR MORE POINTS (Y/N) ? Y
 NUMBER OF POINTS TO BE USED? 8
 AT X =  4.90393  NORMAL STRESS =   310.394
 AT X =  4.15735  NORMAL STRESS =   386.569
 AT X =  2.77785  NORMAL STRESS =   501.288
 AT X =  .975451  NORMAL STRESS =   587.522
 AT X = -.975452  NORMAL STRESS =   587.522
 AT X = -2.77785  NORMAL STRESS =   501.288
 AT X = -4.15735  NORMAL STRESS =   386.569
 AT X = -4.90393  NORMAL STRESS =   310.394
 STRESS INTENSITY FACTOR KI [/ SQRT(PI A)] = 446.443

 REPEAT FOR MORE POINTS (Y/N) ? N

Ready
```

Clearly just a few integration points are needed for this smooth function. The actual stress intensity factor is

$$K_r = 446 \times \sqrt{\pi a} = 79 MPa\, m^{1/2} \tag{6.5}$$

The total stress intensity factor, again using superposition, is to be less than or equal to the fracture toughness hence

$$K_I = K_r + K_{nom} \leq 110 \tag{6.6}$$

and therefore K_{nom}, the stress intensity factor due to some nominal applied stress is

$$K_{nom} = K_{Ic} - K_r = 110 - 79 = 31 = \sigma_{nom}\sqrt{\pi a} \tag{6.7}$$

The additional tensile stress σ_{nom} which will raise the stress intensity factor to the critical value K_{Ic} is thus $118.5 MPa$.

6.5 Notches and fatigue crack growth

In Chapter 2 we introduced the idea that notches and cut-outs can give rise to local stress levels which may be much higher than the applied stress. For example, a simple circular hole can increase the local stress to 3 or more times the applied stress (stress concentration factor $K_t \geq 3$).

Because cracks so frequently start from notches, brittle materials which often break due to crack growth are sometimes referred to as notch sensitive. A conservative approximation to the stress intensity factor at a notch can be simply obtained by multiplying the stress concentration factor by $1.12\sqrt{\pi a}$ where the factor 1.12 is used as a free edge correction (section 3.3). The stress concentration factor may be obtained from tables or from the formula for the stress at the end of an elliptic hole

$$K_t = 1 + 2\sqrt{a/\rho} \tag{6.8}$$

where ρ is the radius at the ends of the ellipse.

Another simple approach to crack growth from notches was proposed by Smith (1977). The notch and crack are represented by a crack of equivalent length L as in figure 6.4. If the actual crack length a is more than $0.13\sqrt{D\rho}$, then any increase in the applied stress due to the notch is neglected and the equivalent length L of the crack is simply taken to be

$$L = D + a \tag{6.9}$$

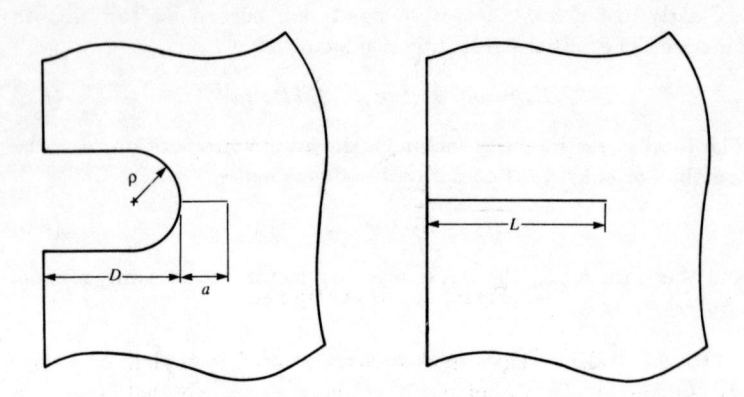

Figure 6.4 Cracked notch and equivalent crack

If the crack length a is less than $0.13\sqrt{D\rho}$ then

$$L = a(1 + 7.69\sqrt{D\rho}) \quad (6.10)$$

Knowing the equivalent length of crack and the nominal applied stress, the stress intensity factor can be estimated, so that for an edge crack (strictly in an infinite sheet)

$$\begin{aligned} K_I &= 1.12\sigma\sqrt{\pi L} \\ &= 1.12\sigma\sqrt{\pi a(1 + 7.69\sqrt{D/\rho})} \end{aligned} \quad (6.11)$$

where 1.12 is the free-edge correction. Since we are discussing fatigue growth we should strictly be using $\Delta\sigma$ and ΔK but for simplicity these quantities are written as σ and K in this section. Approximation by an equivalent length crack seems to work fairly well, agreeing with the results of fatigue tests to $\pm 30\%$ for a wide range of notch size and shape. In effect $\sigma\sqrt{\pi a}$ is being multiplied by a concentration factor like K_t but one which is smaller (and hence more accurate) for longer cracks.

In practice, cracks growing from sharp notches can exhibit some unusual features. Very short cracks grow much more quickly than might be expected from simple elastic analysis, but then the root of the notch is typically in a plastic state due to the very high stress levels there (Figure 6.5). As the crack becomes longer the growth rate may decrease as the crack tip is no longer in the plastic region.

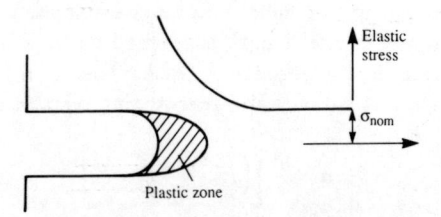

Figure 6.5 Notch tip fields

The equivalent length approach can be used to explain another of the interesting features of very sharp notches, namely that cracks may start at the notch and then stop after growing for a short distance. Neuber proposed a stress distribution ahead of the notch of the form

$$\sigma = \sigma_{nom} K_t \left(\frac{\rho}{\rho + 4x}\right)^{1/2} \quad (6.12)$$

where x is the distance from the notch tip and σ_{nom} is the nominal applied stress. Now when x is small and K_t is large, the stress may be high enough to initiate a fatigue crack if the effective stress intensity factor K is greater than the minimum required for fatigue crack growth K_o.

As the effective crack length increases so K increases (assuming constant stress). However, if the effective length L is large compared to the radius of the notch tip, then K may not increase as quickly as the notch stress field decreases. At some point the notch stress field may drop below that required for continued fatigue growth and the crack simply stops growing.

Now a crack will only grow by fatigue if its stress intensity factor is greater than the threshold value K_0. For an edge crack with effective length L just to grow we must have:

$$K = 1.12\sigma_{nom}\sqrt{\pi L} = K_0 \quad (6.13)$$

so that the stress which just causes growth is

$$\sigma_{nom} = \frac{K_0}{1.12\sqrt{\pi L}} \quad (6.14)$$

and as a good approximation for very sharp notches we can neglect a and take $L = D$.

A more accurate approach for blunter notches is to obtain the fatigue crack length from Neuber's formula using the experimental observation that the critical length is reached when $\sigma = \sigma_0$ where σ_0 is the fatigue stress in the absence of cracks, the plain fatigue strength. Thus, writing a instead of x and rearranging equation (6.12) gives:

$$a = \frac{\rho}{4}\left[\left(K_t\frac{\sigma_{nom}}{\sigma_0}\right)^2 - 1\right] \tag{6.15}$$

The crack length a can now be used to correct the effective crack length L in equation (6.14) as before. This iterative procedure is incorporated into the following simple program.

Program 6.5 NOTCH: Minimum stress for fatigue crack growth from notches

```
100 REM MIN STRESS FOR FATIGUE CRACK GROWTH (MN, METRES)
110 INPUT " NOTCH DEPTH D, RADIUS RHO ";D,R
120 INPUT "THRESHOLD K, THRESHOLD S";KO,SO
130 P1=4*ATN(1.0)
140 C=1+2*SQRT(D/R)
150 L=D
160 S2=0
170 S1=S2
180 S2=KO/(1.12*SQRT(P1*L))
190 PRINT "MINIMUM STRESS";S2
200 A=(R/4)*((C*S2/SO)**2-1)
210 IF A<0.13*SQRT(D/R) GOTO 240
220 L=D+A
230 GOTO 250
240 L=D+A*(1+7.69*SQRT(D*R))
250 IF (ABS((S2-S1)/S2))>0.01 GOTO 170
260 END
```

Sample run

RUN NOTCH

 NOTCH DEPTH D, RADIUS RHO ? .005,.00125
THRESHOLD K, THRESHOLD S? 6.5
? 200

```
MINIMUM STRESS 46.3058
MINIMUM STRESS 45.8121
MINIMUM STRESS 45.8528
Ready
```

Program notes

(1) The stress concentration factor K_t has been estimated in LINE 140 as the stress at the end of an ellipse but could, of course, be input separately if desired.
(2) From LINE 250 the program repeats automatically unless the relative change in stress is less than 1%.
(3) The correction to the initial approximation is very small in this case and could be neglected but the program does also demonstrate how to use the effective length concept.
(4) A notch of 5mm depth and 1.25mm tip radius may start cracking but the cracks are unlikely to continue to grow unless the stress is more than about $46MPa$.

6.6 Experimental data

Before commencing any data analysis, the first thing to consider is the accuracy of the data. If the data are very accurate then it is simplest to fit a function directly through the data points, perhaps based on polynomials as in Lagrange interpolation, or perhaps a different set of functions. If the data contain random experimental errors then a least squares technique is probably best and a simple method is described in section(6.6.1).

One of the commonest fitting problems in fracture mechanics arises in the analysis of crack growth data and so we will consider such a sample problem here. Suppose that a fatigue test has been performed and that the crack lengths measured after a given number of cycles are as follows:

crack length (mm)	10.0	10.7	12.4	14.7	18.2	22.4	29.4
Number of cycles ($\times 10^3$)	5	10	15	20	25	30	35

This data is shown graphically in figure 6.6. The Paris crack growth

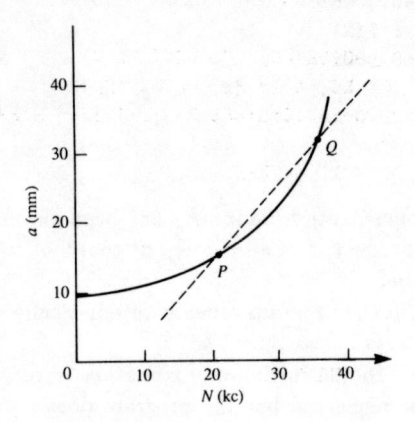

Figure 6.6 Fatigue crack growth rate

law was discussed in chapter 5 and takes the form:

$$da/dN = C(\Delta K)^m \qquad (6.16)$$

where m is called the *Paris exponent* and C is a constant. In order to obtain da/dN we must estimate the gradient of the curve in figure 6.6. Now the gradient of the straight line between the two points P and Q must match the gradient of the curve somewhere between P and Q and, on average, we can take the point to be halfway so that the gradient is

$$da/dN \simeq \frac{a_Q - a_P}{(N_Q - N_P)} \qquad (6.17)$$

at $a = (a_P + a_Q)/2$. The accuracy of this approximation will improve as P and Q get closer together.

If we simply plotted the data for da/dN and ΔK we would obtain a curve. However, taking logarithms of equation (6.16) gives:

$$\log(da/dN) = m \, \log(\Delta K) + \log C \qquad (6.18)$$

which is in the straight line form $y = mx + c$. Hence a graph of $\log(da/dN)$ against $\log(\Delta K)$ should be a straight line of gradient m and intercept $\log C$.

The program below implements the differentiation scheme for the a and N values listed as data and outputs the appropriate logarithms.

Experimental data 135

Program 6.6 : DAN - Gives $\log(da/dN)$, $\log(\Delta K)$ from a, N data.

```
100 REM ** FINDS LOG(DA/DN) AND LOG(DEL K) FROM A,N
110 REM INPUT N IN KC, A IN MM, OUTPUT MM/C
120 DEF FNQ(R)=29.6-185.5*R+655.7*R^2-1017*R^3+638.9*R^4
130 P1=4*ATN(1.0)
140 REM S IN N, B,W IN METRES
150 READ S,N,B,W
160 READ N1,A1
170 N1=N1*1000
180 FOR I=2 TO N
190 READ N2,A2
200 N2=N2*1000
210 L1=LOG((A2-A1)/(N2-N1))
220 A=(A1+A2)/2000
230 R=A/W
240 K=S*SQRT(A)*FNQ(R)/(B*W*10^6)
250 PRINT " K(MPA M^1/2)=";K," A(MM)=";A*1000
260 L2=LOG(K)
270 PRINT " LOG(K)=";L2," LOG(DA/DN)=";L1
280 A1=A2
290 N1=N2
300 NEXT I
310 REM STRESS, NO OF DATA POINTS
320 DATA 36000,7,0.025,0.08
330 REM CRACK GROWTH DATA A1,N1,A2,N2 ETC
340 DATA 5,10,10,10.7,15,12.4,20,14.7
350 DATA 25,18.2,30,22.4,35,29.4
360 END
```

Ready

Sample run

RUN DAN

```
  K(MPA M^1/2)= 26.6493         A(MM)= 10.35
  LOG(K)= 3.28276               LOG(DA/DN)=-8.87387
  K(MPA M^1/2)= 26.5081         A(MM)= 11.55
  LOG(K)= 3.27745               LOG(DA/DN)=-7.98657
  K(MPA M^1/2)= 26.3498         A(MM)= 13.55
  LOG(K)= 3.27146               LOG(DA/DN)=-7.68428
  K(MPA M^1/2)= 26.5051         A(MM)= 16.45
```

```
   LOG(K)= 3.27734              LOG(DA/DN)=-7.26443
   K(MPA M^1/2)= 27.6507         A(MM)= 20.3
   LOG(K)= 3.31965              LOG(DA/DN)=-7.08211
   K(MPA M^1/2)= 31.2263         A(MM)= 25.9
   LOG(K)= 3.44126              LOG(DA/DN)=-6.57128
Ready
```

Program notes

(1) In order to find ΔK for a range of crack lengths we must know Q or K as a suitable function of a. In this case the formula for a compact tension specimen is used in line 120.

(2) There are mixed units in the program which is somewhat confusing, but conventionally da/dN is in mm/cycle and ΔK in $MPa\ m^{1/2}$. The length A in line 220 is changed to metres for the calculation of ΔK.

6.6.1 The least squares method

This method is commonly used for data with experimental errors, ideally with a normal (random error) distribution. The data is approximated by a line which passes near the points in such a way as to minimise the sum of the squares of the deviations from the line. The easiest line to fit to the data is a straight line although, in principle, any curve could be used. The restriction to a straight line is not quite as limiting as it may appear, since many curves may be reduced to a straight line using logarithms as in our current example.

If we start with N pairs of data points x_i, y_i; $i = 1, 2, ..., N$ then it may be shown that the least squares straight line has gradient m:

$$m = \frac{N\Sigma x_i y_i - \Sigma x_i\ \Sigma y_i}{N\Sigma x_i^2 - (\Sigma x_i)^2} \qquad (6.19)$$

and intercept c:

$$c = \frac{\Sigma y_i - m\ \Sigma x_i}{N} \qquad (6.20)$$

where Σ is understood to represent the sum from 1 to N terms. Also, the above least square line assumes that the x-values are relatively accurate and the y-values include any significant errors.

Once the line is fitted then the sum of the squares of deviations of the data can be used to indicate the scatter of the population of data. A measure of scatter can be useful in assessing the variability of test material or even the reliability of experimental techniques. In

Experimental data

statistical terminology an unbiased estimate of the standard deviation (or scatter) of the data is:

$$D = \left[\frac{\Sigma(y_i - mx_1 - c)^2}{N-2}\right]^{1/2} \tag{6.21}$$

The program below calculates and prints m, c and the estimated standard deviation.

Program 6.6.1 : LSQ - Finds a least square straight line fit

```
100 REM ** LSQ - LEAST SQUARE STRAIGHT LINE FIT
110 S=0
120 T=0
130 U=0
140 V=0
150 READ N
160 FOR I=1 TO N
170 READ X,Y
180 S=S+X
190 T=T+Y
200 U=U+X*X
210 V=V+X*Y
220 NEXT I
230 M=(N*V-S*T)/(N*U-S*S)
240 C=(T-M*S)/N
250 PRINT "GRADIENT";M,"INTERCEPT";C
260 REM ****** FIND RMS DEVIATION
270 RESTORE
280 READ N
290 D=0
300 FOR I=1 TO N
310 READ X,Y
320 D=D+(Y-M*X-C)^2
330 NEXT I
340 D=SQRT(D/(N-2))
350 PRINT "ESTIMATED STANDARD DEVIATION";D
360 DATA 6
370 DATA 3.283,-8.874,3.277,-7.987,3.271,-7.684
380 DATA 3.277,-7.264,3.320,-7.082,3.441,-6.571
390 END
```

Ready

Sample run

```
RUN LSQ

GRADIENT 8.26714            INTERCEPT-34.9536
ESTIMATED STANDARD DEVIATION .65887
Ready
```

Program notes
(1) The input data in lines 370 and 380 is taken from program 6.6 above. The two programs could be easily combined into one if both were used frequently together.
The constants in the growth law are thus $m = 8.27$ and $\log C = -34.95$ or $C = 6.60 \times 10^{-16}$.

6.7 Failure assessment diagrams

Failure assessment diagrams are used in an attempt to simplify failure calculations by entering parameters onto pre-prepared diagrams which already incorporate some calculations and/or experimental results. The diagram gives a failure and a 'safe' region and the entered parameters define a point lying in one of these regions. An example for the case of linear elasticity is given in section 5.5 where the values of K_I and K_{II} define a 'safe' region for mixed-mode loading. More difficult problems arise in elastic-plastic deformation and we describe one approach in more detail here. The aim of the 'two-criteria' or R6 assessment procedure is to produce a simple, unified method of describing failure, not simply for fast fracture but for fracture with increasing amounts of plasticity as far as complete plastic collapse. The procedure was developed at the CEGB (R6 stems from a report number) with a view to improving the analysis of pressure vessels.

We have seen in chapter 4 that the size of plastic zones is controlled by the stress state and hence by the size and shape of components, as well as loading, temperature, material, etc. This means that a single material type may behave with varying degrees of plasticity according to its design application. If a material is behaving perfectly elastically, then failure occurs if $K_I \geq K_{Ic}$. If a material is behaving perfectly plastically, then it fails if the stress exceeds the yield stress σ_Y. Most materials exhibit behaviour which is a mixture of these two extremes.

Failure assessment diagrams

The Dugdale strip yield model describes plastic behaviour in the crack tip zone and is outlined in section 4.2, although only a first order approximation to crack tip opening displacement (or COD) is given there. (COD is the opening displacement at the notional tip of a crack caused by plastic deformation. It is thus zero for purely elastic behaviour) The Dugdale model can provide a good approximation over the whole range of elastic-plastic behaviour. To obtain the limiting values which just cause failure, it is assumed that the remote stress causing fracture is σ_c and the plastic collapse stress is σ_Y. Dugdale's formula for the crack opening displacement, δ_c which just gives failure is then given by:

$$\delta_c = \frac{8a\sigma_Y}{\pi E} \ln\left[\sec\left(\frac{\pi \sigma_c}{2\sigma_Y}\right)\right] \quad (6.22)$$

From equation (2.10), the first order approximation to equation (6.22) we have

$$\delta_c = \frac{K_{Ic}^2}{E\sigma_Y} \quad (6.23)$$

Substituting into equation (6.22) and also using $K = \sigma\sqrt{\pi a}$ leads to the formula:

$$K_R = S_R \left(\frac{8}{\pi^2} \ln\left[\sec \frac{\pi}{2} S_R\right]\right)^{-1/2} \quad (6.24)$$

where

$$K_R = K_I/K_{Ic}$$

and

$$S_R = \frac{\sigma}{\sigma_{pf}}$$

where σ_{pf} is the stress required for plastic failure, which may be calculated from σ_Y the yield strength, σ_u the ultimate tensile strength or an average flow stress of $(\sigma_Y + \sigma_u)/2$.

A plot of this equation is shown in figure 6.7. To use the diagram, the maximum likely value of K_I is estimated using any of the techniques already discussed. The minimum value of K_{Ic} is found from data tables or by experiment and the K_R value formed will then be an upper (conservative) bound. Estimating S_R may be difficult, but limit analysis, slip line field theory, or finite element analysis could be used if no relevant data is to hand. Note that the plastic failure stress should take into account the reduction in cross-sectional area due to the presence of a crack. The values of K_R and S_R can then be multiplied by any desired safety factors and plotted onto the R6 diagram, failure being predicted if the point lies outside the curve.

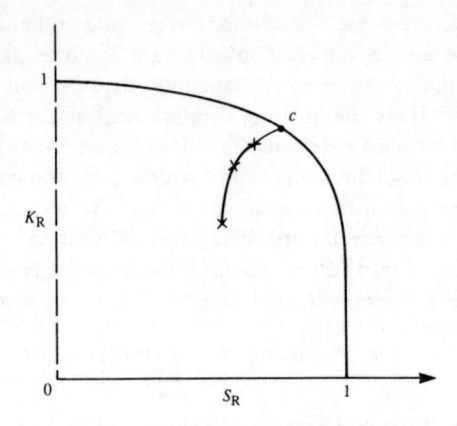

Figure 6.7 R6 failure assessment diagram

A more precise strategy is to estimate K_R and S_R for several defect sizes and plot these on the R6 diagram as shown. The critical defect size can then be estimated by interpolation. The failure assessment diagram is becoming more sophisticated as J-type integrals (introduced in sections 3.6 and 4.6.2) are used to account for work hardening and the effects of secondary stresses are taken into consideration (Chell and Milne 1983).

Another type of failure assessment diagram is the Welding Institute design curve. This curve relates the local strain at the crack site in the uncracked body to the crack opening displacement (COD) and is therefore sometimes known as the COD method (Burdekin and Dawes 1971). Stress and strain concentrations are taken to be equivalent so that they can be estimated from existing notch data etc.. The design curve has been obtained using a large number of tests on thick cracked plates and can give good estimates of failure COD for large welded structures.

Example 6.7

A table of K_R and S_R values for a cracked component is given below, Use the R6 diagram to find the critical crack length at which failure occurs.

Crack length a	K_R	S_R
5mm	0.50	0.58
10mm	0.707	0.62
12mm	0.775	0.68

The three (K_R, S_R) points are plotted on the diagram in figure 6.7. The curve formed is extended to the point where it touches the failure envelope, point c which has the coordinates $K_R = 0.83$, $S_R = 0.80$. Using program 3.2 to interpolate K_R (as x) and crack length a (as y) gives the critical crack length as 13.7mm at $K_R = 0.83$.

6.8 Three-dimensional cracks

Many crack problems are not well represented by two-dimensional approximations. One three-dimensional form which has been extensively studied (because it often arises in practice) is the flat elliptical or semi-elliptical crack. An example of a semi elliptical surface crack is shown in figure 6.8. The curved crack front introduces an additional parameter, namely the crack shape which is represented by the ratio a/c. One effect of the changing shape is a possible variation in the stress intensity factor around the crack front. When the crack is first formed, a/c is often rather small but the stress intensity factor is usually highest at the centre of the crack ($\varphi = 90°$) and the crack grows fastest in this direction (figure 6.8). In the absence of other effects (such as a nearby boundary) then the crack will tend toward a circular shape with $a/c = 1$.

Irwin derived an expression for a completely buried elliptical crack with no nearby boundaries. Assuming a constant remote stress σ then

$$K_I = \frac{\sigma\sqrt{\pi a}}{\Phi}(\sin^2\varphi + \frac{a^2}{c^2}\cos^2\varphi)^{1/4} \qquad (6.25)$$

where Φ is the second kind elliptic integral

$$\Phi = \int_0^{\pi/2} \sqrt{1 - \sin^2\alpha \sin^2\varphi}\, d\varphi \qquad (6.26)$$

and $\sin^2\alpha = (c^2 - a^2)/c^2$. Using a series expansion (Ewalds and Wanhill 1984) Φ can be approximated to an accuracy of about 5% by:

$$h = 1 - (a/c)^2 \qquad (6.27)$$

(a) Semi-elliptical crack (b) Crack growth

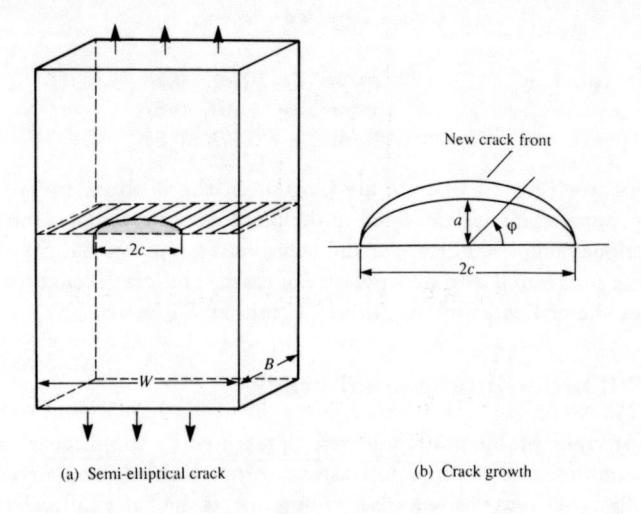

Figure 6.8 The semi-elliptical surface crack

$$m = \left(1 - \sqrt{1 - h^2}\right)/h^2$$
$$\Phi = \frac{\pi}{2(1+m)}\left[1 + \frac{m^2}{4}\right]$$

When elliptical cracks intersect a free surface (figure 6.8 and figure 6.9), then K should be multiplied by a correction factor as is usual in two-dimensions: 1.12 for a simple surface crack and a higher correction factor of 1.2 for a crack at the corner of a block which intersects two surfaces.

If $c = a$ the crack is completely circular and equation (6.25) for a constant tensile stress σ reduces to

$$K_I = \frac{2}{\pi}\sigma\sqrt{\pi a} \tag{6.28}$$

and K is now constant all around the crack front.

The stress intensity factor is affected by all nearby surfaces including those which are not intersected by the crack but this problem can only be solved using numerical techniques at present. Stress intensity factors obtained using a finite element method are given by Raju

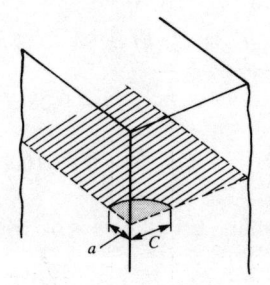

Figure 6.9 A quarter-elliptical surface crack

and Newman (1979) for the crack configuration shown in figure 6.8. Assuming that K is given by

$$K_I = C\sigma\frac{\sqrt{\pi a}}{\Phi} \qquad (6.29)$$

they gave tables of values for the constant C. In the following table values of C are quoted for $\varphi = 90°$.

	\multicolumn{4}{c}{a/B}			
a/c	0.2	0.4	0.6	0.8
0.2	1.173	1.359	1.642	1.851
0.4	1.138	1.225	1.370	1.447
0.6	1.110	1.145	1.230	1.264
1.0	1.049	1.062	1.107	1.112

Having obtained stress intensity factors for a range of a/c and a/B it was then possible to fit an accurate approximating function. The function is rather complex but is given here for the case of constant normal applied stress σ:

$$C = [C_1 + C_2(a/B)^2 + C_3(a/B)^4] C_4 V_1 W_1$$

$$C_1 = 1.13 - 0.09(a/c)$$

$$C_2 = -0.54 + 0.89/(0.2 + a/c)$$

$$C_3 = 0.5 - 1.0/(0.65 + a/c) + 14(1.0 - a/c)^{24}$$

$$C_4 = 1 + [0.1 + 0.35(a/B)^2](1 - \sin\varphi)^2$$

$$V_1 = \left[\sin^2\varphi + (a/c)^2 \cos^2\varphi\right]^{1/4}$$

$$W_1 = \left[\sec\left(\pi c\sqrt{a/B}/W\right)\right]^{1/2}$$

Example 6.8

A pressure vessel of diameter $R = 500mm$ and wall thickness $B = 4mm$ has to withstand a pressure of $P = 10MPa$. Cracks of up to $10mm$ in length and about $1mm$ deep are discovered in the walls. If the fracture toughness of the material is $60MPa\ m^{1/2}$ what action should be taken ?.

The important step is clearly to estimate a stress intensity factor for this configuration. The hoop stress σ is given by:

$$\sigma = PR/B = 10 \times 10^6 \times 0.25/0.004 = 625MPa \qquad (6.30)$$

Since the radius of curvature of the vessel is large compared to the wall thickness, we can assume a flate plate approximation. However, as well as the hoop stress there is also a pressure on the crack surfaces of $10MPa$ making the total stress $635MPa$.

The crudest method is to use a two-dimensional approximation and assume a plane strain crack of depth $1mm$ in a $4mm$ strip. Since the assumption of plane strain implies that the crack is infinitely long, this K-value may be expected to be an overestimate. From Rooke and Cartwright (1976) $Q = 1.50$ and hence

$$K_I = \sigma Q\sqrt{\pi a} = 635 \times 1.5\sqrt{0.001\pi} = 53.4 MPa\ m^{1/2} \qquad (6.31)$$

Alternatively, considered as an elliptical crack we have $a/c = 0.2$ and $a/B = 0.25$. Using Lagrange interpolation on the tabled values of C gives

```
RUN LAG

LAGRANGE INTERPOLATION
NUMBER OF POINTS TO BE GIVEN? 3
XI,YI? 0.2,1.173
XI,YI? 0.4,1.359
XI,YI? 0.6,1.642
COEFFICIENT OF X** 0    IS  1.084
COEFFICIENT OF X** 1    IS  .202502
COEFFICIENT OF X** 2    IS  1.2125
NEW X VALUE? 0.25
AT X = .25    Y = 1.21041

ANOTHER X-VALUE (Y/N)? n
Ready
```

and using equation (6.29) with $C = 1.21$ and $\Phi = 1.016$ gives $K = 43.8 MPa m^{1/2}$.

Another alternative is to use equation (6.29) with the approximating function for C. This is not elementary as a hand calculation and it is better to use a BASIC program as below.

Program 6.8 ELL: Stress intensity factors for a semi-elliptic surface crack.

```
100 REM *** FINDS K FOR ELLIPTIC CRACK, UNITS MN, METRE
110 INPUT "PLATE WIDTH W, THICKNESS B";W,B
120 INPUT "HALF CRACK WIDTH C, DEPTH A";C,A
130 INPUT "APPLIED STRESS S";S
140 P1=4*ATN(1.0)
150 V=P1/2
160 A6=A/C
170 A8=A/B
180 PRINT " A/C=";A6," A/B=";A8," C/B=";C/B
190 H=1-A6^2
200 M=(1-SQRT(1-H^2))/H^2
210 P=P1*(1+M^2/4)/(2*(1+M))
220 C1=1.13 - 0.09*A6
230 C2=-0.54 + 0.89/(0.2 + A6)
240 C3=0.5 - 1.0/(0.65 +A6) + 14*(1.0 -A6)^2
250 C4=1+(0.1 + 0.35*A8^2)*(1-SIN(V))^2
260 V1=(SIN(V)*SIN(V) + A6^2*COS(V)*COS(V))^(1/4)
```

```
270 W1=SQRT(1/COS(P1*C*SQRT(A8)/W))
280 C0=(C1+C2*A8^2 + C3*A8^4)*C4*V1*W1
290 PRINT "   C=";C0
300 K=C0*S*SQRT(P1*A)/P
310 PRINT "   K =";K
320 END
```

Sample run

```
RUN ELL

PLATE WIDTH W, THICKNESS B? 1,.004
HALF CRACK WIDTH C, DEPTH A? .005,.001
APPLIED STRESS S? 635
  A/C= .2        A/B= .25      C/B= 1.25
   C= 1.24969
   K= 43.7604

Ready
```

Program notes

(1) Since the plate width is unspecified it is simply taken to be a relatively large number ($W = 1$) which should have little effect on the stress intensity factor obtained.

(2) The value of K agrees closely with the value obtained by interpolation from the table. However, the program has two advantages; one, it is faster and easier to use, and two, it may be used for a wider range of values than specified in the table. Using interpolation to go outside the data table is rather more uncertain.

We have obtained three estimates of the stress intensity factor, the first, a two-dimensional approximation, is expected to be conservative. The other two, based on the same original data should be similar and are effectively identical. Since our best estimate of K is within 30% of the given fracture toughness and since the crack depth is uncertain, the vessel could fail at any time and it should not be used again.

6.9 References

Burdekin, F.M. and Dawes, M.G. (1971) Practical Application of Fracture Mechanics to Pressure Vessel Technology, 28, Institution of Mechanical Engineers, London.

Chell, G.G. and Milne, I. (1983) STP No. 803, 11-179 Philadelphia, ASTM.

Ewalds, H.L. and Wanhill R.J.H. (1985) *Fracture mechanics*, Edward Arnold, London.

Kanninen, M.F. and Popelar, C.H. (1985) *Advanced Fracture Mechanics*, Oxford University Press.

Parker, A.P. (1981) *The Mechanics of Fracture and Fatigue*, E & F Spon Ltd, London.

Raju, I.S. and Newman, J.C. (1979) Stress intensity factors for a wide range of semi-elliptical surface cracks in finite thickness plates. *Engg. Frac. Mech.* 11, pp 817-829.

Rooke, D.P. and Cartwright, D.J. (1976) *A Compendium of Stress Intensity Factors*, HMSO London.

Smith, R.A. (1977) A simplified method of predicting the rates of growth of cracks initiated at notches. in *Fracture Mechanics in Engineering Practice* P. Stanley (Ed.), Applied Science Publishers, pp173-182.

PROBLEMS

(6.1) A cracked component has a width $W = 600mm$ and is loaded with an cyclic stress of between 80 and $140MPa$. If $K_{Ic} = 115MPa\ m^{1/2}$, what is the critical crack length ? You may assume that the following data are available.

a/W	0.02	0.09	0.12
Q	1.34	1.57	1.65

If the material has a Paris exponent of 3.4 with $C = 4.5 \times 10^{-12}$ what is the expected lifetime of the component ?

(6.2) A crack emanates from the edge of a circular hole in a large sheet of material loaded by a stress of $300MPa$ normal to the crack. What is the critical crack length if the material has a fracture toughness of $55MPa\ m^{1/2}$? Is your estimate conservative ?

(6.3) A component has been treated so that a compressive stress of $10MPa$ exists in a $1mm$ thick surface layer. The stress then changes linearly for the next $1mm$ to a tensile stress of $10MPa$. Calculate the stress intensity factor for a number of crack lengths up to, say, $5mm$ and plot the results.

(6.4) A notch is $4mm$ deep and has a tip radius of $0.7mm$. What is the minimum cyclic stress required for cracks initiated at the notch to continue growing if the material has $\Delta K_0 = 10MPa\ m^{1/2}$ and a plain fatigue limit of $230MPa$?

If the applied stress is $50MPa$ and a crack of length $0.5mm$ exists what is its effective stress intensity factor ?

(6.5) A fatigue test uses a compact tension specimen to obtain the following results.

Crack length (mm)	12.4	12.6	13.2	13.8	14.1	14.7	16.1
Number of cycles ($\times 10^3$)	5	10	15	20	25	30	35

Determine a Paris crack growth law for this material.

(6.6) A cracked vessel with a $6mm$ crack is estimated to have a stress intensity factor of $K_I = 74.25MPa\ m^{1/2}$ and a stress to plastic collapse stress ratio $S_R = 0.80$. If, for a $10mm$ crack $K_I = 95.7MPa\ m^{1/2}$ and $S_R = 0.80$, estimate the critical crack length given that the minimum fracture toughness of the material is $110MPa\ m^{1/2}$.

(6.7) A cylindrical pressure vessel is $700mm$ in diameter and $750mm$ high with walls 6mm thick. The material used has a fracture toughness of $67MPa\ m^{1/2}$. If it is to contain a pressure of $14MPa$ what is the critical crack size ?. What aspect ratio would a semi-elliptic crack have to possess if the vessel is required to leak before total fracture ?

Index

approximations using superposition, 39–43
assignment statement, 2
BASIC, 1–11
    functions, 2
    statements, 10
    variants, 10
bend specimen, 89
boundary element method, 44–46
brittle failure, 80

checking programs, 9
circular crack, 142
circular hole, 40, 45
COD, 139
COD method, 140
compact tension specimen, 81, 91
conditional operators, 5
conditional statements, 5
configuration correction factor, 33
constant amplitude cyclic loading, 94
contour integral, 57, 90
crack closure, 33
crack growth, 93
    by fatigue, 94–100
    direction, 105–113, 115
    energy, 19
    unstable, 21
crack growth rate, 94, 99

crack opening displacement, 51, 55, 105
crack opening modes, 25
crack path
    stable, 114
    unstable, 113
crack resistance force, 21
crack shape, 72
crack sizing, 123
crack sliding displacement, 104, 105
crack tip
    displacement expansions, 24
    displacements, 101
    plasticity, 69–91
    stress expansions, 24
    stresses, 101
crack tip opening displacement (COD), 88–89
crack tip singularity, 71
crack tip stress, 40
critical crack length, 20, 123–126
critical strain energy density, 114
critical stress, 113, 114
critical tensile stress, 114
cyclic loading, 93, 94

damage-tolerant, 121
DATA, 4
DEF FN, 9
design strategies, 121–123

DIM, 7
displacement extrapolation, 54
ductile shear failure, 79
Dugdale, 71
Dugdale model, 139
dye penetrant, 123

edge crack, 40, 63, 98, 124, 130
effective crack length, 72
elastic-plastic deformation, 89, 138
elasticity formulae, 13
elements, 45, 46
elements for crack problems, 47
ellipse, 129, 133
elliptic integral, 141
elliptical crack, 122, 141–146
energy
    and crack growth, 19
energy release rate, 21, 57
equivalent length crack, 129
experimental errors, 136

fail-safe, 121, 122
failure
    brittle, 12
    ductile, 12
failure assessment diagrams, 138–141
failure load, 86
fast fracture, 113–116, 123
fatigue, 93, 116
fatigue crack growth, 94–100, 129–133
fatigue strength, 132
fatigue test, 133
finite element method, 44, 46–47, 90, 142
flow stress, 139
FOR .. NEXT, 6
fracture toughness, 80, 113, 123

free surface correction, 142
functions
    argument of, 2
    defined, 9
    trigonometric, 2

Gauss quadrature, 59
Gauss-Chebyshev quadrature, 59, 63
GO TO, 6
GOSUB, 8
gradient of curve, 134
Green's function, 128

Heat treatment, 126
hierarchy of operators, 2
Hooke's law, 13, 14, 16
hoop stress, 144

IF THEN, 5
INPUT, 3
integration, 40
interactive programs, 3
interpolation, 34–37, 146
interpolation functions, 45
Irwin model, 94
isoparametric, 45

J-integral, 57, 89–91
J-integrals, 140

Lagrange interpolation, 34, 35, 45, 124, 144
leak-before-break, 122
least squares, 136–138
load-displacement curves, 82
loops, 6

magnetic particle, 123
maximum stress, 40
maximum tensile stress, 106–112, 114

INDEX

mean stress, 40
mesh, 45, 46
mixed-mode, 27, 53, 105, 114
mixed-mode loading, 101, 105
modes of crack opening, 25
modulus of rigidity, 14
multiple assignments, 10

Neuber, 131
NEXT, 6
non-destructive testing, 123
normal distribution, 136
notches, 129

ON .. GOSUB, 9
ON .. THEN, 8
output, 4
overlapping crack surfaces, 39
overload, 100

Paris
    equation, 96
    exponent, 96
    law, 99, 134
path-independent integral, 90
plane strain, 16, 27, 78, 144
plane strain fracture toughness, 80–88
plane stress, 15, 27, 78
plastic collapse, 138
plastic failure, 70
plastic zone, 94, 138
    shape, 74–78
    size, 70–74
        Dugdale model, 71
        Irwin model, 71
Poisson's ratio, 13, 110
polynomial, 35, 44
potential energy, 89
precracking, 81, 86
pressure vessel, 138, 144

principle of superposition, 27, 38
PRINT, 4

quadratic element, 47
quarter-point elements, 48, 54

R-values, 99–100
R6 procedure, 138
radiography, 123
READ, 4
REM, 8
residual stress, 89, 100, 126–129
RESTORE, 4
RETURN, 8
reversed yielding, 94
Rice's J-integral, 57
run-time input, 3

safe-life, 121
second order terms, 101–105, 116
semi-elliptical crack, 141
shear lip, 80
shear modulus, 14
shear stress, 70, 79
short cracks, 130
shot peening, 126
Simpson's rule, 40, 96
stable crack growth, 88, 89
standard deviation, 137
STEP, 6
stiffener, 122
STOP, 6
strain, 13
    'engineering' shear, 14
    axial, 13
    shear, 14
    three-dimensional, 14
strain energy, 17, 19, 61

strain energy density, 57, 62, 90, 106–112, 114
strain energy release rate, 89
stress
    direct, 13
    shear, 14
    three-dimensional, 14
    transformation, 16
stress concentration, 18
    on ellipse, 18
stress concentration factor, 18, 43, 129, 133
stress corrosion, 99
stress intensity factor, 24, 33–66
    range, 95
    threshold, 95
stress range, 100
stress relief, 127
stress transformation, 106
string, 8
subroutines, 8
subscripted variables, 7
superposition, 27, 38–43, 61
surface energy, 20
surface tension, 20

TAB, 10
tabulation, 5
tensile stress, 106
testpiece design, 85–88
three point bend specimen, 81, 91
three-dimensional cracks, 141–146
threshold K, 95, 131
traction-singular elements, 48
two criteria, 138
two-dimensional approximation, 144
types of fracture, 79–80

ultimate tensile strength, 139
ultrasonic, 123

variable amplitude loading, 100
virtual crack extension, 56
Von Mises criterion, 79

welded structures, 89, 140
welding, 126
Welding Institute, 140

yield criterion, 70
yield stress, 70
Youngs Modulus, 13